W9-CKF-390

ENGINEERING DESIGN GRAPHICS

SKETCHING, MODELING, AND VISUALIZATION

James M. Leake

Department of Industrial & Enterprise Systems Engineering
University of Illinois at Urbana-Champaign

with special contributions by

Jacob L. Borgerson

Underwriters Laboratories Inc. Northbrook, Illinois

WILEY John Wiley & Sons, Inc.

James M. Leake
dedicates this work to
Richard Costello Leake
and
Kathyrn Alice Duff

Jacob L. Borgerson
dedicates this work to
Erin

EXECUTIVE PUBLISHER	Don Fowley
ASSOCIATE PUBLISHER	Dan Sayre
ACQUISITIONS EDITOR	Michael McDonald
EDITORIAL ASSISTANT	Rachael Leblond
SENIOR PRODUCTION EDITOR	Ken Santor
TEXT AND COVER DESIGNER	Madelyn Lesure

This book was set in 3B2 by Thomson Digital Inc. and printed and bound by R.R. Donnelley—Crawfordsville. The cover was printed by Phoenix Color Corporation.

This book is printed on acid free paper. ⊗

To order books or for customer service please, call 1-800-CALL WILEY (225-5945).

ISBN 978-0-471-76268-3

Printed in the United States of America

10 9 8 7 6 5 4 3 2 1

☐ PREFACE

The traditional first-year engineering graphics course has undergone significant change in the past quarter century. While the emergence of computer-aided design (CAD) and the expansion of the graphics curriculum to include design are perhaps the most significant developments, more recent trends include a movement away from 2D CAD and towards 3D parametric solid modeling, an increased emphasis on freehand sketching at the expense of instrument drawing, and a greater focus on the development of spatial visualization skills. All of this has occurred despite a strong countervailing trend to de-emphasize graphics in order to accommodate other material in the four-year undergraduate engineering curriculum.

The aim of this book, then, is to provide a clear, concise treatment of the essential topics included in a modern engineering design graphics course. Projection theory provides the instructional framework, and freehand sketching the means for learning the important graphical concepts at the core of this work. The book includes several hundred sketching problems, all serving to develop the student's ability to use sketching for ideation and communication, as well as a means to develop critical visualization skills.

Engineering design serves to bracket the graphical content of the book, with an introductory chapter on the engineering design process, and a concluding chapter on reverse engineering and redesign. Material contained in the first chapter is based upon introductory material found in leading engineering design textbooks. This chapter concludes with concept design project descriptions, intended primarily for execution by student teams using parametric modeling software. Thumbnail images of student projects are also included. Using commercial products, the final chapter describes the reverse engineering process, including product dissection, documentation, analysis, and improvement.

A chapter on computer-aided product design software, with an emphasis on parametric solid modeling, is also included. The chapter is designed to complement, rather than to replace, instructional materials for a specific CAD package. The chapter provides an overview of different kinds of CAD software, general modeling concepts shared by all parametric modelers, and a discussion of some downstream applications of CAD models, including 3D printing.

Two additional chapters, provided as appendices, are also included. The first appendix is on perspective projections and sketching. This material has been included as it reflects the way that engineering graphics has traditionally been taught at the University of Illinois at Urbana-Champaign (UIUC), starting from the general (e.g., perspective projection), and moving to the specific (e.g., multiview orthographic projection). The second appendix chapter is on geometric dimensioning and tolerancing (GD&T), and has been included in response to many reviewer requests.

Key features of the book include the following:

- A succinct, scaled-down approach, with important concepts distilled to their essence
- Hundreds of sketching problems to help students learn the language of technical graphics, and to develop their sketching, visualization and modeling skills
- Assembly problems requiring a wide range of modeling tools, not just extrusions and revolutions
- Lots of visualization materials: sections on multiview visualization and the section view construction process are included, as are missing view problems, problems that require students to mentally rotate and then sketch a different pictorial view of the object, problems that require students to find a partial auxiliary, missing, and pictorial view when two views are given, as well as section view problems
- A strong student focus, with many examples showing what students can produce in an engineering design graphics course
- A chapter on engineering design that reflects the thinking of leading engineering design educators
- A chapter on reverse engineering, something unique to engineering design graphics textbooks
- A chapter on GD&T, a topic not typically found in abbreviated engineering graphics texts

- A unified planar projection theory framework that provides a common basis for understanding the relationships between different kinds of sketches (i.e., perspective, oblique, isometric, multiview), and also serves as an introduction to the study of computer graphics
- Several detailed multi-step example sketching problems that provide students with problem solving procedural templates

The premise and framework for the book was largely conceived in conversations with the book's original editor at Wiley, Joe Hayton. Much of the book's content, in particular Chapters 2 through 6, 8, and Appendix A, is strongly influenced by a system of teaching engineering graphics that has developed over the years in the Department of General Engineering at UIUC. In particular, I would like to acknowledge the work of my immediate predecessor, Michael H. Pleck. Hallmarks of this approach include a focus on planar projection theory, starting from general case perspective projections and advancing to more specific projection types, as well as an emphasis on spatial visualization problems. A special thanks goes out to the "best and brightest" in the State of Illinois, that is, UIUC engineering students who have made significant contributions to the content of this work. The book's co-author, Jacob Borgerson, is responsible for the many fine problems that are included in the book's end of chapter exercises, as well as for his careful reading of and thoughtful comments on the text. Our thanks go out to the book's many reviewers, including Ghodrat Karami, North Dakota State University; Ken Youssefi, University of California, Berkeley/San Jose State University; Patrick McCuistion, Ohio University; Andrea Giorgioni, New Jersey Institute of Technology; Jeff Raquet, University of North Carolina—Charlotte; Ramarathnam Narasimhan, University of Miami, College of Engineering; Michael Keefe, University of Delaware; Robert D. Knecht, Colorado School of Mines; and Randy Emert, Clemson University.

The inspiration for certain chapters deserves special mention. Chapter 1 on engineering design is based on the introductory chapters from some of best books on engineering design, including those of G. Pahl and W. Beitz, George Dieter, Rudolph Eggert, and Clive Dym. Chapter 9 on Product Dissection, Reverse Engineering and Redesign, is largely based on the work of Sheri Sheppard, Kevin Otto and Kristin Wood, and Ronald Barr. Appendix B, Geometric Dimensioning and Tolerancing, is inspired by the article on GD&T on the eFunda Web site www.efunda.com, and by the excellent work on this subject by Max Raisor.

Thanks also to the students listed below, whose artwork appears in the Team Concept Design Projects thumbnail figures found in the Questions section at the end of Chapter 1: Joseph Gonzalez, Matt Gelber, BJ Dori, Kevin Hanecamp, Sean Conway, Kevin Burrus, Peter Stynoski, Brian Budd, Daniel Manhart, Timothy Schulz, Don Schwer, Buder Shageer, Andy Block, Brian Ferencak, Kevin Knapp, Scott Quinlan, Jarrod Waldeck, Kristin Althoff, Ken Dollaske, Phillip Martorana, Brian Matesic, Ji-young Shin, Paul Dworzanski, Derek Evans, Kristen Hoebbel, Jared Surratt, Dan Tatje, Jon Schmid, Beau Hill, John Dunlap, Jason Wiltz, Aaron Koronkowski, Beth Richter, Casey Roth, Dan Schafman, Kyle Hanstad, Shivani Kapadia, Emmalyn Riley, Nathan Schroeder, Brandon Wynn, Matt Ralph, Garrett Leffelman, Gerardo Rangel, Joseph Hodskins, Hanna Albrecht, Liz Chapman, Myles Hathcock, Emma Kress-Israel, Robin Palmissano, Dominic Menoni, Michael Rybalko, Stephanie Schachtrup, Dan Weidner, Bob Wille, Eric Biesen, Clinton Clark, Anthony Mareno, Lauren Stromberg, Paul Keutelian, Jon Adam, Daniel Anderson, Jordan Sneed, Kyle Hewerdine, Chris Ruback, Justin Gantenberg, Matt Howard, Travis Brown, Jae Cha, Eric Leigh, Rick Riley, Jennifer Lee, Jamie Hung, DJ Lee, Eric Mueller, Daniel Cholst, Frank Lam, Laurence Lin, Madison Major, Gordon Yang, Jaed Daum, Kyle Farver, Erik Johnson, Jessica Wayer, Aaron D'Souza, Greg Schebler, Sebastian Witkowski, Jack Xu, Joseph Cassidy, Jeremy Mlekush, Benny Poon, Brian Pruis, Luke DeTolve, Tom Boerman, Joel Ingram, Jin Soo Ohm, Clems Thermidor, Ted Yelton, Brian Davids, Sebastian Derza, Jake Metzger, Brett Zitny, Eric James, Frank Lam, Spenser Murakami, Narayan Patel, Tasia Bradley, Colin Das, Jon Homan, Dino Hopic, Adam Tate.

James M. Leake
Urbana, Illinois November 2007

□ CONTENTS

CHAPTER

ENGINEERING DESIGN

☐ INTRODUCTION

Design is the central activity of the engineering profession. *Engineering design* can be defined as a set of decision-making processes and activities that are used to determine the form of a product, component, system, or process, given the functions desired by the customer.[1] The term *function* refers to the behavior of the design; that is, what does the design need to do? *Form*, on the other hand, has to do with the appearance of the design. A product's form refers to its size, shape, and configuration, as well as the materials and manufacturing processes used to produce it.

Engineering design is a part of the larger *product realization process*. As we see in Figure 1-1, product realization starts with a customer need and ends with a finished product that satisfies this need. The product realization process consists of design and manufacturing processes that are used to convert information, materials, and energy into a completed product. The stages of the product realization process include sales and marketing, industrial design, engineering

design, production design, manufacturing, distribution, service, and disposal. *Product development* refers to the first stages of the product realization process up to manufacturing. Product development includes engineering design, as well as sales/marketing, industrial design, and production design.

☐ ASPECTS OF ENGINEERING DESIGN

There are many different aspects comprising our notion of engineering design. For instance, it is a *process*, one that prominently involves both *problem-solving* and *decision-making* activities. Engineering design also employs both *analysis* and *synthesis*. By nature it is *interdisciplinary* and *iterative*. Even prior to the emergence of such modern concepts as *concurrent engineering* and the design team, there has always been a strong *social* aspect to engineering design. Finally, in keeping with the main topic of this book, engineering design is characterized by strong *graphical* elements.

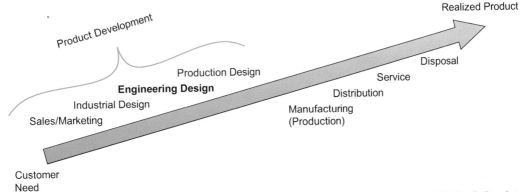

Figure 1-1   Product realization process (Eggert, Rudolph, J., *Engineering Design*, 1st Edition, © 2005, Pg. 3. Reprinted by permission of Pearson Education, Inc., Upper Saddle River, NJ.)

[1] Rudolph J. Eggert, *Engineering Design*, Pearson Prentice Hall, 2005.

Engineering design is really about solving problems. In fact, a simple definition of engineering design is "a structured problem-solving approach."[2] Concisely described, the design process is no more than identifying a problem, carefully researching and defining the problem in order to better understand it, creatively generating possible alternative solutions to address the problem, evaluating these candidate solutions to ensure their feasibility, making a rational decision, and then implementing it. One of the hidden merits of an engineering education is that this problem-solving framework becomes so ingrained that it can easily be adapted to deal with life's many problems, technical or otherwise.

Although the need to make decisions is apparent throughout the course of a design's evolution, decision making in conjunction with engineering design typically refers to that part of the design process where competing feasible solutions are evaluated and an optimal solution is decided upon. Because of the numerous **trade-offs** involved, these types of decisions are often difficult to make. Trade-off examples include strength versus weight, cost versus performance, or towing power versus free running speed. By optimizing one criterion, the optimum position of the other criterion is often sacrificed. Decision making is still something of an art, requiring solid information, good advice, considerable experience and sound judgment. In recent decades, however, a mathematically based **decision theory** has been developed that is commonly used in the development of commercial products.

Synthesis is the process of combining different ideas, influences, or objects into a new unified whole. From this perspective, engineering design can be viewed as a synthesis technique, one used for creating new products based on customer needs. More specifically, synthesis refers to creative approaches used to generate potential solutions to a design problem.

Analysis, on the other hand, refers to the process of breaking a problem down into distinct components in order to better understand it. From an engineering design standpoint, analysis often refers to tools used to predict the behavior and performance of potential solutions to a design problem. Good design requires both divergent thinking (i.e., synthesis), which is used to expand the design space, and convergent thinking (i.e., analysis), which is used to narrow the design space by focusing on finding the best alternatives in order to converge to an optimal solution.

As a synthetic process, engineering design is interdisciplinary in nature. While relying heavily on basic science, mathematics, and the engineering sciences, engineering design still retains its artistic roots. Eugene Ferguson, in his book *Engineering and the Mind's Eye*, notes that "many of the cumulative decisions that establish a product's design are not based in science." He points out that while design engineers certainly make decisions based upon analytical calculations, many important design decisions are based on engineering intuition, a sense of fitness, and personal preference.[3]

An important element of engineering design is communication. It is often noted that a significant portion of an engineer's time is spent in communication (oral, written, listening) with others. Add to this the fact that modern engineering design draws on decision making, optimization, engineering economy, planning, applied statistics, materials selection and processing, and manufacturing, and the interdisciplinary nature of engineering design should be clear.

Although engineering design is frequently described as a sequential process that moves from one stage to the next, it can also be portrayed as a design spiral, where each stage is visited more than once (see Figure 1-2). In any case, feedback loops, like that shown in Figure 1-3,[4] are built into each design process step. Based on the latest information, it is often necessary to iterate the various design process steps.

---

[2] Arvid Eide et al., *Engineering Fundamentals and Problem Solving*, McGraw-Hill, 1997.

[3] Eugene Ferguson, *Engineering and the Mind's Eye*, MIT Press, 1997.

[4] George Dieter, *Engineering Design: A Materials and Processing Approach*, McGraw-Hill, 1991.

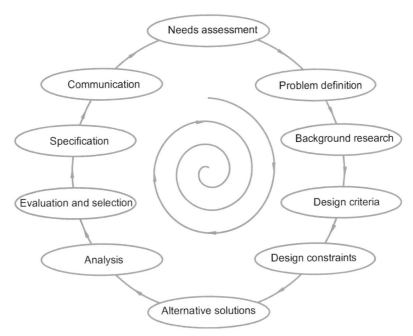

Figure 1-2 Engineering design process

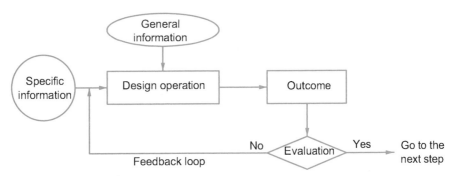

Figure 1-3 Design process feedback loop

As we have already mentioned, the actual engineering design process is not nearly as structured as that described in engineering design textbooks. There has always been a strong social aspect to the design process, perhaps best captured in the notion of the napkin or back-of-the-envelope sketch.[5] In fact, a reasonable definition of the engineering design process is "a social process that identifies a need, defines a problem, and specifies a plan that enables others to manufacture the solutions."[6] This social aspect of engineering design is further reinforced with the emphasis in the recent decades on project design teams. Figure 1-4

[5] For example, the original concept for the Seattle Space Needle was sketched on a placemat in a Seattle coffee house.

[6] Karl Smith, *Project Management and Teamwork*, McGraw-Hill, 2000.

shows two engineers holding a discussion over an engineering drawing.

Figure 1-4 Engineers discussing a design project (Courtesy of Jensen Maritime Consultants, Inc.)

Graphics is used throughout the entire engineering design process. Freehand sketching is useful both for generating and for communicating ideas. Computer-aided design (CAD) is used to evaluate design alternatives and perform *what-if analyses*. CAD models can also be used in downstream analysis, prototyping, and manufacturing applications, as well as for marketing and sales. Finally, CAD models and drawings are used for traditional documentation purposes.

☐ ANALYSIS AND DESIGN

The undergraduate engineering curriculum includes both analysis and design content. Engineering analysis is characterized by breaking a complex problem up into manageable pieces. *Modeling*, for example, is an analysis technique used to simplify a more complicated real-world situation. In a typical analysis problem, the input data is given and there is a single solution to the problem.

Design problems employ synthesis as well as analysis. Design problems are open-ended; that is, there is more than one solution. The correct answer is unknown. For this reason optimization and decision-making techniques are often employed in association with design. Design problems are typically constrained by time, money

and legal issues; they are multidisciplinary, and of undefined scope.

☐ PRODUCT ANATOMY

A *product* is a designed object or artifact that is purchased and used as a unit.[7] Products vary in complexity depending upon the number, type, and function of their components. A *component* is either a part or a subassembly. A *part* is a piece that requires no assembly, while an *assembly* is a collection of two or more parts. A *subassembly* is an assembly that is included in another assembly or subassembly. For example, a bicycle is a product that can also be thought of as an assembly composed of subassemblies and parts. A rear hub is a subassembly on a bicycle, while a handlebar is a part on a bicycle. Both the hub and the handlebar are bicycle components.

☐ DESIGN PHASES

Product design projects, as conducted by large manufacturing companies like Boeing or Motorola, pass through several distinct phases (see Figure 1-5). In the *formulation* phase information

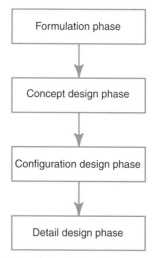

Figure 1-5 Engineering design phases

---

[7] John Dixon and Corrado Poli, *Engineering Design and Design for Manufacturing: A Structured Approach*, Field Stone Publishing, 1999.

is gathered in order fully to understand the design problem. Functional (customer, company) requirements of the product, as well as constraints and evaluation criteria, are identified; finally engineering performance targets are developed. In the *concept design* phase the product is decomposed initially by component (form) and then by function in order to understand it better. Concept alternatives are then generated to meet each product subfunction; these subfunction alternative concepts are then analyzed to determine their feasibility, and finally evaluated based upon the evaluation criteria. The concept design phase concludes with the selection of the best concept alternatives for each product subfunction. By the conclusion of the *configuration design* phase, the type and number of components of the product, their arrangement, and their relative dimensions have been determined. In the *detail design* phase, a package of information that includes drawings and specifications sufficient to manufacture the product is prepared.

DESIGN PROCESS OVERVIEW

The design phases describe the actual procedure taken by a manufacturing company in order to bring new products to market. The engineering design process, on the other hand, is more an instructional framework used to describe the idealized steps taken by an engineering team in developing a design.

The engineering design literature includes many different versions of this process, each with a different number of stages or steps. The design process template (see Figure 1-2 on page 3) employed in this work[8] includes 1) needs assessment, where specific needs (problems) are identified, 2) problem definition, where the problem is clarified, 3) background research, where research is conducted to gather additional information about the problem, 4) design criteria, where desirable characteristics of the design are identified, 5) design constraints, where quantitative boundaries

that limit the possible solutions to the problem are identified, 6) alternative solutions, where candidate solutions are conceptualized and generated, 7) analysis, where the alternative solutions are analyzed in an effort to determine their feasibility, 8) evaluation and selection, where the remaining solutions are evaluated with respect to a weighted criteria, and the best candidate is selected, 9) specification, where drawings and technical specifications documenting the design are created, and finally, 10) communication, where written reports and oral presentations describing the design are developed. Many of the steps (e.g., problem definition) may be revisited multiple times as additional information comes to light. Further, the stages may not necessarily be performed in the order specified here, and some steps may be omitted entirely.

NEEDS ASSESSMENT

The engineering design process starts with the recognition of a need that can potentially be satisfied using technology. In many cases needs are not identified by engineers, but rather by the general public. Potential clients approach design firms with a specific need. In a large manufacturing company, the sales or marketing department maintains contact with its customer base and identifies many of these needs. In a more systematic approach, product ideas are generated by a product planning group.

Needs arise for a variety of reasons, including 1) a product redesign in order to make it more profitable or effective, 2) the establishment of a new product line, 3) to protect public health and safety, or to improve quality of life, 4) an invention, often by an individual, that is then commercialized, 5) opportunities created by new technology or scientific advances, 6) a change in rules, requirements, etc. The outcome of the needs assessment is a list of needs or requirements that then becomes a part of the problem definition.

PROBLEM DEFINITION

Once a list of needs have been developed, the next step in the design process is clearly and

[8] Adapted from Eide.

carefully to formulate the problem to be solved. At first glance this step may seem somewhat trivial, but the true nature of the problem is not always obvious. A poorly defined problem can lead to a solution search that is either misdirected or too limited in scope. A good problem definition should focus on the desired functional behavior of the solution, rather than on a specific solution. For example, if the issue is lawn maintenance, then "Design a better lawn mower" presupposes the solution. If the problem is restated as "Design an effective means of maintaining lawns," then our options are kept open.[9] By defining the problem as broadly as possible, novel or unconventional solutions are less likely to be overlooked. Another tip when formulating a problem is to focus on the source of the problem, rather than the symptoms. In any event, a complete definition of the problem should include a formal **problem statement**.

A problem statement attempts to capture the essence of a problem in one or two sentences. A problem is characterized by three components: 1) an undesirable initial state, 2) a desired goal state, and 3) obstacles that prevent going from the undesired to the desired state.[10] A good initial problem statement should consequently describe the nature of the problem, as well as express what the design is intended to accomplish. In further iterations, the obstacles that prevent reaching the desired state may also be included.

☐ BACKGROUND RESEARCH

Having broadly defined the problem, the next step in the engineering design process is research. Background research is conducted in order to obtain a deeper understanding of the problem. Information is sought regarding the target user, the intended operating environment, additional constraints that bound the solution, prior design

solutions, etc. Useful questions that may be posed at this stage include: what has been written on the topic, what must the design do, what features or attributes should the solution have, is something already on the market that may solve the problem, what is right or wrong with how it is being done, who markets the current solution, how much does it cost, how can it be improved upon, will people pay for a better one if it costs more?

Common background research sources include existing solutions, the library, the internet, trade journals, government documents, professional organizations, vendor catalogues, experts in the field, etc. Regarding existing solutions, **reverse engineering** is the process of physically disassembling an existing product to learn how each component contributes to the product's overall performance. Manufacturing companies use reverse engineering techniques in order to gain insight into a competitor's design solutions. Applied systematically, **best-in-class** solutions can then be identified for a broad range of common engineering problems. Reverse engineering is discussed in further detail in Chapter 9.

☐ DESIGN CRITERIA

Design criteria[11] are desirable characteristics of the solution. They tend to originate from experience, research, marketing studies, customer preferences, and client needs. Design criteria are used at a later stage in the design process to qualitatively judge alternative design solutions. Design criteria may be categorized as either general or specific. General criteria categories applicable to almost all design projects include cost, safety, environmental protection, public acceptance, reliability, performance, ease of operation, ease of maintenance, use of standard components, appearance, compatibility, durability, etc. Criteria categories specific to a certain design might include weight, size, shape, power, physical requirements for use, reaction time required for

---

[9] Restating the problem in this way led to the invention of the Weed Eater® lawn trimmer.

[10] G. Pahl and W. Beitz, *Engineering Design: A Systematic Approach*, Springer-Verlag London, 1996.

[11] In engineering design literature, design criteria are frequently expressed as design objectives or design goals.

Table 1-1 **Tugboat design criteria**

Specific	General
Bollard pull (thrust, towing power)	Cost
Maneuverability	Ease of operation
Speed (free running)	Ease of maintenance
Stable work platform	Reliability
Visibility from the pilothouse	Durability
Stability (safety)	Use of standard components
Habitability	Environmental protection
Seakeeping	Appearance
Workmanship	

operation, noise level, etc. Table 1-1 provides examples of both general and specific criteria for a tugboat design similar to that shown in Figure 1-6.

Figure 1-6 Tugboat photograph (Courtesy of Jensen Maritime Consultants, Inc.)

□ DESIGN CONSTRAINTS

Design constraints limit the possible number of solutions to a given design problem. Design constraints are quantitative boundaries associated with each design objective. They establish maximum, minimum or permissible ranges for physical or operational properties of the design, environmental conditions affecting the design

or impacting the environment, ergonomic requirements, as well as economic or legal constraints. A feasible design solution must satisfy all of the design constraints.

Design constraint categories include: 1) physical-space, weight, material, etc., 2) functional or operational-vibration limits, operating times, speed, etc., 3) environmental–temperature ranges, noise limits, effect upon other people, 4) economic – cost of existing competitive solutions in the market place, 5) legal–governmental and other regulations, and 6) ergonomics/human factors– strength, intelligence, anatomical dimensions, etc. Some example design constraints, organized by category, for a tugboat are shown in Table 1-2 on page 8.

□ ALTERNATIVE SOLUTIONS

Having thoroughly researched the problem, the next step is systematically to generate as many alternative product solutions as possible for subsequent analysis, evaluation and selection. The most common technique used for generating ideas is ***brainstorming***. The objective of brainstorming is to develop as many ideas as possible in a limited amount of time. Roughly half an hour is typically allotted for a brainstorming session. Emphasis is placed on the quantity rather than the quality of the ideas. Free expression is essential; the ideas can be evaluated at a later time. On occasion one

Table 1-2 Tugboat design constraints (sample)

Physical	Environmental
Length overall (LOA) < 100 feet	Engine room noise level < 120 decibels
Fuel > 30,000 gallons	
Fresh water > 6,000 gallons	**Legal (Regulatory Bodies)**
ASTM A-36 steel used for hull	American Bureau of Shipping (ABS)
Heel less than 1/4 degree	American Society of Mechanical Engineers (ASME)
Railing height > 39 inches	American Society of Testing & Materials (ASTM)
	Environmental Protection Agency (EPA)
Functional/Operational	International Maritime Organization (IMO)
Bollard pull > 30 tons	Occupational Safety and Health Administration (OSHA)
Horsepower > 2500 BHP	
Running speed > 12 knots	**Ergonomic**
Accommodations for 6 crew	Accommodations headroom > 6 feet
Economic	
Cost < $8 million	

group member's seemingly impractical notion serves as the inspiration for a teammate's more viable idea. It is important that all group members participate as equals. A group leader can start with a clear statement of the problem, invite ideas regarding the solution, and set the pace of the session. A recorder is also necessary, in order to make note of everything said.

As we discussed in the section on product anatomy, a product is composed of different components. Each component has its own form, as does the product. Similarly, a product has a primary function, but it also has secondary or subfunctions. For example, the primary function of a coffee maker is to make coffee, but it also needs to store ground coffee and water, hold a filter, brew coffee, convert electricity to heat, and keep the coffee warm. To ensure a robust design, multiple concept alternatives need to be generated for each product subfunction.

A *morphological chart* is a tool that can be used to expand the number of concept alternatives. Product subfunctions are listed versus concept alternatives in a table. Subfunctions are listed vertically and the corresponding concept alternatives horizontally in a row. Figure 1-7 shows a morphological chart for a portable water filter. Using this technique, a vast number of possible solutions can be identified. Although some of these solutions may be incompatible, morphological

Function	Solution 1	Solution 2	Solution 3	Solution 4	Solution 5
draw water	manual	hose	hose w/ float	hose w/ float, prefilter	
move water	hand pump	lever pump	gear pump	dual piston pump	gravity feed
filter water	ceramic	ceramic w/ carbon core	fiberglass	labyrinth	iodine resin
collect water	manual	threaded adaptor base	built-in collection bottle		

Figure 1-7 Morphological chart for a portable water filter

analysis is nonetheless useful in identifying a great many combinations, some of which may be unusual or innovative.

In order to compete globally, large manufacturing companies engaged in product realization systematically optimize a product's design at the

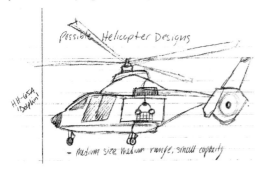

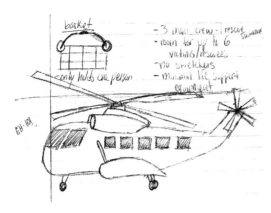

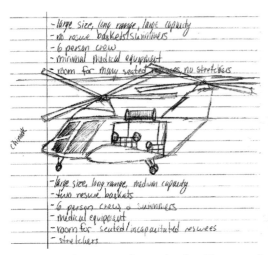

Figure 1-8 Alternative solution sketches (Courtesy of Matthew Patton.)

subfunction level. For first-year student design projects, though, it may be sufficient to generate, evaluate and then select from product level alternative solutions. For example, each member of the design team may propose an alternative solution to the problem, as shown in Figure 1-8.

Many designs too, have distinct solutions that have developed over time. Moveable bridges, for example, include single and twin span bascule drawbridges, vertical lift and swing bridges (see Figure 1-9 on page 10), each one serving as an alternative solution to the moveable bridge problem.

Students may well come up with their own innovative solutions to these time honored alternatives, as seen in the student drawbridge concept design solution in Figure 1-10 on page 10.

□ ANALYSIS

The concept alternatives generated in the previous stage must be analyzed and then evaluated in order to determine which concepts are worth pursuing. It is possible that some of the candidate concepts that have been generated thus far are not actually feasible. Design analysis, which uses mathematical and engineering principles to evaluate the performance of a solution, may be used to verify the functionality and manufacturability of the concept alternatives. The objective is to eliminate any infeasible solutions. Although the candidates are at this point no more than abstract concepts, still it may be possible to eliminate some of them using back-of-the-envelope calculations, investigating their manufacturability, etc.

□ EVALUATION AND SELECTION

After analysis, the remaining feasible candidate concepts are evaluated. Design evaluation is a stage in the design process when results from analyses are assessed to determine which alternative is best. The alternatives are evaluated with respect to weighted evaluation criteria. The evaluation criteria are derived from the design criteria established earlier in the design process, although these criteria may have been modified and further refined based on knowledge gained in the course of the design's development.

(Nick Tsolov/i Stockphoto)

(Jim Jurica/i Stockphoto)

(Lee Foster/Digital Railroad, Inc.)

Figure 1-9 Moveable bridge alternative solutions
(Jim Wark/Peter Arnold, Inc.)

Figure 1-10 Innovative student drawbridge design solution (Courtesy of Yang Cui, Allison Dale, David Shier, Michael Marcinowski)

It is the responsibility of the design team to come to a group decision regarding the relative importance of the different criteria. The evaluation criteria are then weighted in terms of their importance. It is common to have the criteria weights add up to 100. Once the weighting has been established, the concept alternatives (or product subfunctions) can be evaluated with respect to the weighted criteria. Table 1-3 provides an example of the evaluation procedure used by a student group in a first-year engineering design graphics course to design a dune buggy.

An image of the resulting dune buggy concept design appears in Figure 1-11.

Another example is provided in Figure 1-12. In this industry-sponsored project, a senior design team was asked to develop a new and improved endcap display to be used in grocery and other stores. In an effort to evaluate different concepts with respect to the design criteria, the group

Table 1-3 Alternative design evaluation process

Weighted Design Criteria

Cost	30%
Safety	10%
Weight and Power	15%
Durability	15%
Ease of Operation	20%
Simple Construction	10%

Alternative Design Rankings (1–10)

Weighted Design Criteria	Brian A's Drawing	Dan's Drawing	John's Drawing	Brian S's Drawing	Nilay's Drawing
Cost	8	10	9	8	8
Safety	7	8	6	9	10
Weight and Power	10	6	9	10	7
Durability	8	7	7	9	8
Ease of Operation	7	10	9	6	9
Simple Construction	9	7	6	8	8

Alternative Design Results (Rankings Multiplied By Percentages)

Weighted Design Criteria	Brian A's Drawing	Dan's Drawing	John's Drawing	Brian S's Drawing	Nilay's Drawing
Cost	2.4	3	2.7	2.4	2.4
Safety	0.7	0.8	0.6	0.9	1
Weight and Power	1.5	0.9	1.35	1.5	1.05
Durability	1.2	1.05	1.05	1.35	1.2
Ease of Operation	1.4	2	1.8	1.2	1.8
Simple Construction	0.9	0.7	0.6	0.8	0.8
Total	**8.1**	**8.45**	**8.1**	**8.15**	**8.25**

Figure 1-11   Concept dune buggy design (Courtesy of Dan Fey, Nilay Patel, Jack Streinman, Brian Aggen, Brian Shea)

developed the evaluation matrix shown in Figure 1-12. Each column in Figure 1-12 shows a different endcap criterion. These include modularity, adjustable shelving, mobility, futuristic design, ease of cleaning, beverage advancement, packout, and header. In the cells beneath each column, images illustrating different ways to satisfy the specific criterion are shown. Also included in each cell is a rating (the circled number in the lower right-hand corner) assigned by the design team. This rating is used to indicate the performance of the particular solution in meeting the criterion. Using this and other techniques, the

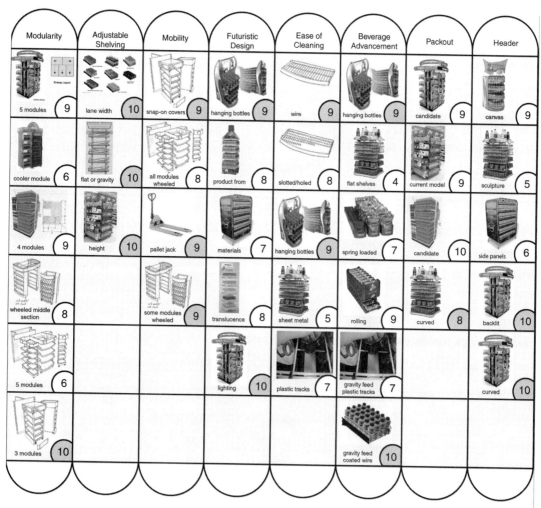

Figure 1-12   Evaluation matrix for an endcap display (Courtesy of Franklin Wire and Display; Jennifer Bessette, Michelle Wentzler, Madison Major, Faye Hellman)

design team produced the endcap design shown in Figure 1-13.

Figure 1-13 Endcap display design (Courtesy of Franklin Wire and Display; Jennifer Bessette, Michelle Wentzler, Madison Major, Faye Hellman)

□ SPECIFICATION

Having selected the best concept alternatives, or simply the best alternative solution, it is now time to configure and arrange the different product features, parts, and components using computer-aided design (CAD). *Configuration design* refers to that part of product development where the number and type of components, parts or geometric features and how they are spatially arranged or interconnected are determined, as well as the approximate relative dimensions of the components, parts and features.

This is typically accomplished using a general arrangement or layout drawing. Figure 1-14 on page 14 shows general arrangement and outboard profile drawings for a 96' tugboat. The tug design employs two rudder propellers mounted at the stern of the vessel. The rudder propellers can be fully rotated through 360 degrees independently of one another, giving the vessel exceptional maneuverability and propulsion char-acteristics.

In addition to the general arrangement drawing, a written specifications document is typically produced at this time. A dictionary definition of *specification* is "a detailed description of a particular thing, especially one detailed enough to provide somebody with the information needed to make that thing." The design specification document is consequently a comprehensive description of the product, including the intended uses, physical dimensions, materials employed, operating conditions, and performance targets, as well as the functional, maintenance, testing, and delivery requirements of the designed object. Relevant codes and regulations to be adhered to in the course of the product's design and manufacture are also spelled out in these specifications. Table 1-4 on page 15 shows an excerpt taken from the written specifications developed for the tugboat shown in Figure 1-14.

These drawings and specifications may then be used in a formal bidding process called a Request for Proposals (RFP). The drawing package would be sent out to different contractors, in this case shipyards, to be competitively bid upon for the right to construct the vessel. Assuming that the RFP process is successful and an agreement is reached to continue the product's development,[12] the project would then move from the configuration to the ***detail***

---

[12] In a manufacturing firm, the decision would be made internally whether to continue the product's development beyond the configuration design phase.

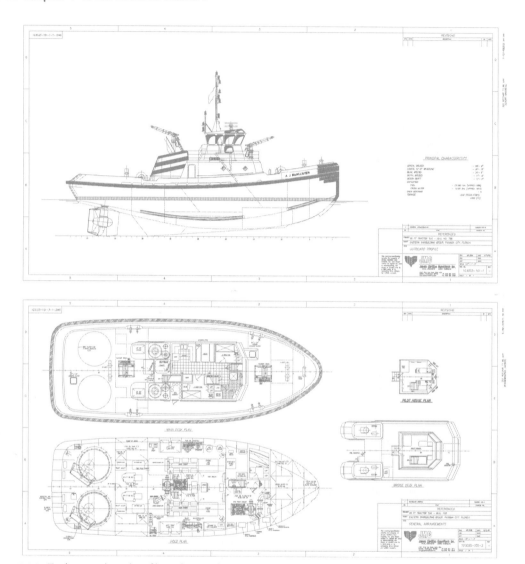

Figure 1-14   Tugboat outboard profile and general arrangement drawings (Courtesy of Jensen Maritime Consultants, Inc.)

design phase. The outcome of the detail design phase is a complete package of information, including many more drawings and more detailed written specifications, sufficient to allow the contractor to fabricate the product.

These drawings are referred to as **working drawings**, also called production or detail drawings. Working drawings are discussed in Chapter 8. Working drawings typically include part detail drawings, assembly drawings (see Figure 1-15 on page 16), and a bill of materials (BOM), or parts list. An assembly drawing with a bill of materials is shown in Figure 1-16.

Note that the term *specification* is also used in engineering design to refer to a target value, performance measure, or benchmark used to evaluate the success of a design. Taken in this sense, specifications begin to emerge early in the design process, as the designer clarifies the client's or company's understanding of the problem, and the potential user's needs. Examples of performance specifications taken from the

Table 1-4 Excerpt from written specifications (Courtesy of Jensen Maritime Consultants, Inc.)

GROUP 100 - HULL STRUCTURE

SECTION 101 – GENERAL INFORMATION

See SECTION 012 for OWNER furnished equipment.

The CONTRACTOR shall design and construct the principal hull structure as described herein. The hull and deckhouse is constructed of steel. All plating shall be ABS Grade A and/or ASTM A-36. All steel shall be wheel-abrated and pre-primed. Steel Certificates shall be provided to the OWNER, along with location of the certified steel.

All workmanship and welding shall be in accordance with the following:

ABS "Rules for Building and Classing Steel Vessels Under 90 meters (295 feet) in Length" (for Hull)

ABS "Rules for Building and Classing Steel Vessels for Service on Rivers and Intracoastal Waterways", Part 3, Section 7, Passenger Vessels (for Superstructure)

Classification Society certification of the structure is not required, but structural calculations supporting compliance with one of the above listed Rules shall be submitted to the OWNER for approval in areas that the CONTRACTOR proposes any changes. Calculations shall show clear references to rules being used.

Construction details shall be in accordance with ABS' "Guide for Shipbuilding and Repair Quality Standards for Hull Structures During Construction", latest edition. Plating shall be fitted fair and free from buckles or uneven sight edges. All formed plates or shapes shall be formed true to the required alignment, shape or curvature. Where flanges are used for attachments, the faying edges shall be beveled and free from hollows. Shims shall not be used to correct improper fit. Members shall be in alignment before welding is undertaken. No fairing compounds shall be used. Warpage or distortion that prevents the installation of the final welded assembly into the boat is not acceptable.

"Panting" or "oil-canning" of any panel in shell, deckhouse or decks is not permitted. Filling compound shall not be used to compensate for unfairness in the boat structure. Every effort shall be exercised to construct a vessel with fair and undistorted surfaces. This shall include diligence for careful fit-up, proper weld sequences and utilizing minimum weldments to achieve required structural strength.

The maximum unfairness between hull, deck and house stiffeners shall be t/2, where "t" is the plate thickness.

Penetrations of deck, bulkhead, and shell plating shall be reinforced as required to maintain structural integrity.

All cuts shall be neatly and accurately made with edges cleaned for welding. All sharp edges exposed to personnel or equipment shall be dressed or ground to avoid injury to operating or maintenance personnel or damage to equipment. Internal corners shall be filleted and external corners shall be rounded off. Ragged edges or sharp projections shall be removed.

All burning of notches and holes in the structure shall be carefully laid out and shall be regular in outline with jagged edges removed and with rounded corners. Semicircular "rat holes" shall be provided in framing members in way of passing welded seams or butts in the attached plating. Notches and rat holes shall be machine cut.

(Continued)

Where it is necessary to provide holes for passage of wiring, piping or ductwork through vessel structure, special penetration details shall be used to maintain structural and fire insulation integrity and oil/water/air tightness as appropriate. All penetrations of fire boundaries shall be collared and insulated on both sides of the fire boundary and as otherwise necessary to meet regulatory requirements. In general, penetrations through bulkheads shall utilize Nelson Firestop 4″ × 6″ transit or equal. All details of such penetrations shall be submitted to the OWNER for approval.

All edges or corners of the bulwarks which could have a fire hose dragged over or around them shall be radiused using 1-1/2 to 2 inch diameter pipe (or equivalent flanged radius).

The CONTRACTOR shall ensure that precautions are taken to prevent rusting of exposed plate and new welds. The OWNER is extremely concerned that excessive rust bloom will compromise the ability of the paint system to obtain a secure hold on the sub-surface, especially in way of the welds. Painting shall be conducted within a controlled environment, protected from the elements, through use of either permanent or portable structures. See SECTION 631 for a description of surface preparation and other coating requirements.

written tugboat specifications document cited in Table 1-4 include:

- *Propulsion system shall consist of a pair of EPA Tier 1 certified, 4-cycle diesel engines with a minimum continuous rating of approximately 2480 BHP.*

- *The boat shall undergo a two-hour endurance trial run at the maximum engine rating. During this trial it shall be demonstrated that all mechanical parts of the propulsion unit and all auxiliaries are in satisfactory operating condition, and that propulsion system steady state*

conditions are within engine manufacturer's tolerances. Inspections shall be carried out for leaks in all piping systems and any structural defects. The readings of all installed gauges and meters shall be recorded at 15-minute intervals.

☐ COMMUNICATION

At this point, the product's development transitions from design to manufacture. The final step in the engineering design process is to

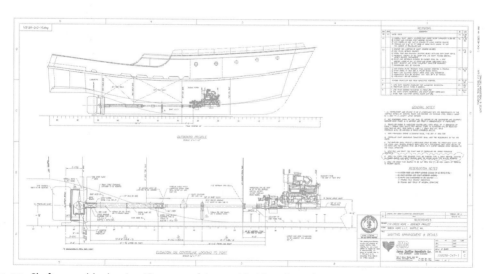

Figure 1-15 Shaft assembly drawing (Courtesy of Jensen Maritime Consultants, Inc.)

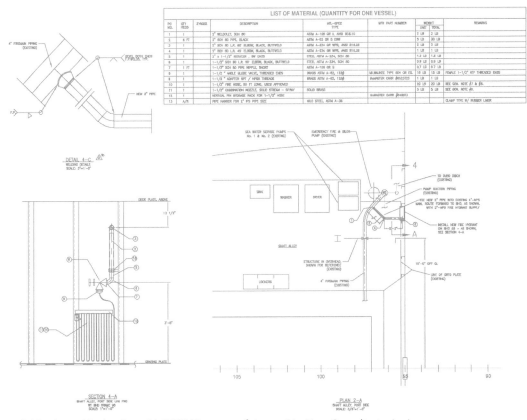

Figure 1-16   Assembly drawing with BOM (Courtesy of Jensen Maritime Consultants, Inc.)

communicate the design using written reports and oral presentations.

Written Reports

Although known more for their math than their verbal skills, most experienced engineers are in fact good writers. Written communication consumes a large portion of an engineer's work week. Written documents composed by engineers include letters, email, memoranda, technical reports, technical papers and proposals. The audience for these technical documents includes other engineers, designers, supervisors, vendors, regulatory agencies, and customers. Regardless of the audience and type of report, technical writing requires a clear, concise expression of what can be very complex issues, concepts, and arguments.

A typical outline for a technical report might include the following sections:

Introduction

Background

Procedure

Results

Discussion (of results)

Conclusions

References

Appendices

Recommended report writing steps

When writing a technical report or paper, it is recommended that the writing process be broken up into several distinct sessions. Try to start each session when you are at your best (e.g., early morning, late afternoon). Once the session begins, try to ensure that you won't be interrupted for at least a few hours. Finally, wait a day before starting the next session. Shown below is a series of session steps that can be used to write papers and technical reports.

1. *Brainstorm ideas*—In the first session, write down everything that you wish to say, making no effort to structure the material. The emphasis should be on the free flow of ideas. This session is probably best done using a pen and paper.

2. *Preliminary outline*—In the second session, attempt to structure the brainstorming output in bullet form. What are the topic headings and subheadings? How are the topics ordered? What material falls under which heading? At this stage transition to a word processor, if you have not already done so.

3. *Detailed outline*—Read the initial outline, revising and amplifying as needed. Move things around, add and delete material. Do the topic headings and bullet items easily transition and flow? Is there sufficient content to make the argument? If not, do more research to fill in gaps. Amplify on the bullet items, adding detail as needed. Move from phrases to complete sentences.

4. *Start writing*—Normally with the introduction, but not necessarily. This is where the true struggle takes place. Progress can be painfully slow; concentration, determination, grit are essential. Aim for coherence and continuity. After a significant struggle, you may have to settle for less. As long as an honest effort is made, a fresh start will almost always help. A good day's work is a couple of pages.

5. *Carefully and critically edit* what has already been written for coherence, continuity, grammar. Revise, rewrite, rearrange.

6. *Critically reread* the document from start to finish, making final editorial changes as necessary.

Oral Presentations

As an engineer, you will occasionally be called upon to present the results of your work. Pre-sentations are made to deliver progress and summary reports, to pitch an idea or proposal, and also to present a paper at a conference or technical society. Whatever the purpose, it is important to carefully prepare for, and then practice, the delivery of the presentation. When designing the presentation, it is important to know your audience so that you can anticipate and fulfill their needs. Good visual aids and graphics are important, as is a well-organized, logical development of the main ideas.

Some oral presentation guidelines include:[13]

1. Speak clearly so that you can be heard at the back of the room.

2. Vary the pitch or tone of voice occasionally.

3. Add enthusiasm to the delivery.

4. Expect to be somewhat nervous. Take a deep breath and relax a moment before beginning.

5. Start on time, stick to the schedule, and finish on time; do not go over.

6. Maintain eye contact with your audience. Pick a few individuals in the room and look at them as you present.

7. Use a pointer when appropriate.

8. Introduce all team members and the project title.

9. Show an overview slide describing the structure of the presentation.

10. Make your slides clear and crisp. Make sure all slides have sufficient contrast to show the text clearly.

11. Make bullets brief, with typically no more than five words.

12. Each bullet slide should require your explanation in order to be well understood by the audience. In this way the audience will rely on you to bring the presentation together.

---

[13] Adapted from the Senior Project Design Manual used in GE 494, Senior Project Design, at the University of Illinois Urbana–Champaign.

13. Photos should be large and clear. Use labels, arrows, etc. to define and point out important features.

14. When showing a graph, briefly define the axes and then state the purpose of the graph and what it is intended to show.

15. Use PowerPoint® transitions sparingly.

16. Be prepared; know the purpose of each slide. Do not include any information on a slide that you cannot explain.

17. In developing your ideas, move from the general to the specific.

18. Distribute any handouts before the presentation begins.

19. Be clear about your conclusions and what they mean. Show how your conclusions and final recommendations satisfy the problem statement.

20. Have a *war chest* of extra slides for the Q&A period. Anticipate questions and prepare extra slides that will help you answer them.

21. In your practice sessions, notice how often you say "Um," "you know," etc. Try to avoid this nervous habit. Silence is best when you have nothing to say.

□ CONCURRENT ENGINEERING

In the face of ever-increasing global competition, it has become necessary for successful manufacturing companies engaged in product development to continuously 1) shorten product development times, 2) improve product quality and performance, and 3) reduce product cost. The modern approach used to accomplish these daunting tasks is called concurrent engineering.

In the years after the Second World War, product development as conducted in large manufacturing companies was essentially a serial process, as shown in Figure 1-17 on the left. The specialized functions of the company, represented as distinct departments, worked in comparative isolation. The free exchange of information across functional boundaries was not encouraged. The most commonly cited example of this behavior occurred at the interface between design and manufacturing, where the

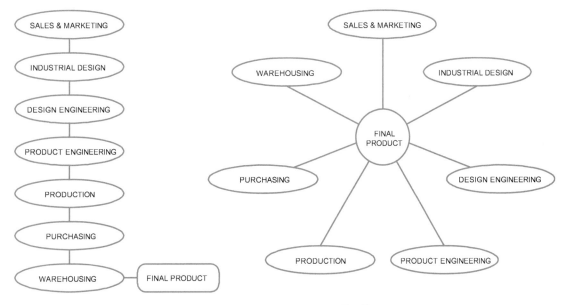

A. Traditional   B. Concurrent

Figure 1-17   Traditional versus concurrent engineering

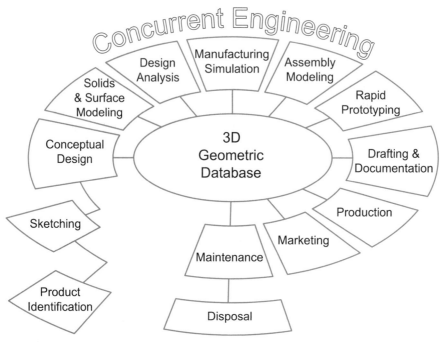

Figure 1-18   Concurrent engineering and the CAD database (Courtesy of Barr, Kreuger, and Juricic)

design group developed a design in isolation and then ***tossed it over the wall*** to manufacturing. Manufacturing was then left to modify the design in order to meet manufacturing process, material, and equipment constraints. Although these changes were costly, a lack of competition made them bearable.

With the emergence, in recent decades, of formidable global competitiveness, these inefficiencies can no longer be tolerated. ***Concurrent engineering*** is a team approach to product design in which team members, representing critical business functions, work together under the coordination of a senior manager. The cross-functional team is typically composed of members from such areas as sales/marketing, industrial design, design engineering, industrial engineering, manufacturing engineering, purchasing, and production. See Figure 1-17 for a graphical comparison of the concurrent versus traditional engineering models.

One of the benefits of concurrent engineering's team approach is the open exchange of

product information. Concurrent engineering is sometimes represented in association with the single most important source of product information, the CAD database, as shown in Figure 1-18.

Although concurrent engineering strives to incorporate all aspects of a product's life cycle (i.e., design, manufacture, distribution, service, and disposal) into the product's development, the central motivation for concurrent engineering is to ensure that manufacturing considerations are taken into account early and throughout the entire course of the design process. In the following section, design for manufacture and assembly, this point is addressed more fully. The chapter concludes with a discussion of another one of concurrent engineering's most important attributes, teamwork.

☐ DESIGN FOR MANUFACTURE AND ASSEMBLY

Design changes occurring late in the design process are especially detrimental to a product's cost, quality, and time to market. Design for Manufacture (DFM) and its cousin, Design for Assembly (DFA), are relatively new areas that strive to minimize these late-occurring design changes by formalizing the relationship between design and manufacture.

While Design for Manufacture aims to improve the fabrication of individual parts, Design for Assembly strives to reduce the time and cost required to assemble a product. Taken together, DFM and DFA practices help ensure that the product and the manufacturing processes used to produce it are designed together. Recently a new term, Design for X (DFX), has become popular. DFX describes any of the various design methods that focus on specific product development concerns. These include Design for the Environment, Design for Reliability, Design for Safety, Design for Quality, etc.

Some of the more important DFMA guidelines include:

- *Minimize the total number of parts*—The fewer the parts, the easier and faster the product's assembly.
- *Minimize part variations*—This can be accomplished, for example, by using the same size fastener throughout.
- *Design parts to be multifunctional*—A single part can be designed to serve more than one purpose, eliminating the need for additional parts.
- *Design parts for multiuse*—Use the same part in different products.
- *Design parts for ease of fabrication*—Near-net-shape manufacturing processes (e.g., injection molding) are best; avoid machining if possible.
- *Design parts with self-fastening features*—Use snap and press-fits whenever possible; avoid screws, bolts, etc.
- *Use modular design*—Design self-contained elements, blocks, or chunks that can be fabricated under ideal conditions, and then connected to the main assembly.

- *Minimize assembly direction*—Design products to be assembled along a single, preferably vertical axis (also called z-axis loading).
- *Minimize handling requirements during assembly*—Positioning parts is costly; design features can be used to simplify positioning; avoid parts that tangle.
- *Maximize compliance in assembly*—Use generous tapers, chamfers, radii; guiding features; use one component as a base.

☐ TEAMWORK

The adoption of concurrent engineering is credited with enhancing productivity in design and manufacture, allowing companies engaged in product development to maintain and even increase market share despite the recent upsurge in global competition. Teamwork is the single most important distinguishing factor of the concurrent engineering philosophy. Teams are increasingly common on projects and in engineering firms. Even without this recent trend, though, engineering design has always been a social activity relying upon teamwork skills like communication, collaboration and cooperation.

A *team* is a group of people with complementary skills and knowledge who work together toward common goals and hold one another mutually accountable. *Teamwork*, on the other hand, refers to a demonstrated attitude and ability to accomplish team goals.

There is of course no assurance that a team will outperform a collection of individuals. Research suggests that there are four identifiable team performance level categories. These include teams that: 1) perform below the level of the average member, 2) don't quite get going but struggle along at or slightly above the level of the average member, 3) perform quite well, and 4) perform at an extraordinary level, where the members are deeply committed to one another's personal growth and success.[14] Research also

[14] Smith, 2000.

Table 1-5 Code of cooperation, Boeing Airplane group, training manual for team members

 1. Every member is responsible for the team's progress and success
 2. Attend all meetings and be on time
 3. Come prepared
 4. Carry out assignments on schedule
 5. Listen and show respect for the contributions of other members; be an active listener
 6. Constructively criticize ideas, not persons
 7. Resolve conflicts constructively
 8. Pay attention; avoid disruptive behavior
 9. Avoid disruptive side conversations
10. Only one person speaks at a time
11. Everyone participates; no one dominates
12. Be succinct; avoid long anecdotes and examples
13. No rank in the room
14. Respect those not present
15. Ask questions when you do not understand
16. Attend to your personal comfort needs at any time, but minimize team disruption
17. Have fun

suggests that sufficient time is necessary for a team to reach the higher performance levels.[15]

Successful teams are characterized by certain traits or skills. These include the use of group norms, as well as communication, leadership, decision making, and conflict management. *Group norms* are guidelines or standards of behavior that are agreed upon by the group. By establishing and adhering to these guidelines, many disruptive conflicts can be avoided. Table 1-5[16] shows an example of group norms.

Group decision making is considerably more difficult than a decision taken individually. There are a number of ways to make a group decision, ranging from decision by authority to consensus. A decision based on authority takes the least amount of time, while decisions based on consensus require more time, but generally result in the best decision. A consensus decision is one in which the team thoughtfully examines all the issues, and agrees upon a course of action that does not compromise any strong convictions of a team member.

Conflicts inevitably arise between team members. A conflict refers to a situation in which the action of one person interferes with the actions of another person. Conflicts can either be constructive or destructive. In a constructive conflict, the source of the disagreement is based on ideas or values, while in a destructive conflict the dispute is based on personality. In order to function properly, it is essential for teams to develop rules that prohibit destructive conflict.

There are a number of strategies for dealing with conflict. These include:

1. *Avoidance*—simply ignore the situation, and hope that the problem resolves itself over time.
2. *Smoothing*—let the other side have his/her way.

[15] Eggert, 2005.
[16] Taken from Smith, 2000.

3. *Forcing*—one side imposes a solution.

4. *Compromise*—meet the other side halfway. While sometimes successful, this approach does not get to the underlying cause of the disagreement.

5. *Constructive engagement*—the two parties strive to get to the heart of the matter, and then work to resolve the situation through negotiation. Only this approach offers the possibility of reaching a satisfactory resolution to a serious conflict.

☐ QUESTIONS

TEAM CONCEPT DESIGN PROJECTS

Working in teams (three to five individuals), develop a virtual model illustrating the configuration (e.g., size, shape, contents) of the concept design projects described below. Deliverables may include a written report(s), oral presentation, design criteria and constraints, candidate solutions, sketches, working drawings, rendered images, animations, 3d prints etc.

Collapsible Skateboard Skateboarding, popu-

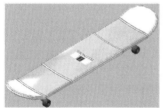

larized in the '70s, has become a classic mode of transportation. Today, boards can be purchased with various dimensions; however, the awkward length of the board often makes storage and transport difficult. The purpose of this project is to develop a skateboard that can collapse from greater than 30 inches in length to less than 12 inches. Its folded height must not exceed 6 inches and its folded width must not exceed 10 inches. The weight of a 250-lb individual should be supported by the design. It must have a handle for easy transport.

Planet Exploration Robot An autonomous planetary or lunar exploration robot should be designed. The mobile robot should be capable of producing the power necessary to complete

its mission, navigate difficult terrain, gather information with sensors, and transmit this information back to earth. The robot should be equipped with cameras, antennas, a rock abrasion tool, remote sensing instruments, a sample acquisition arm, a mass spectrometer, etc. The robot should also be able to produce stereoscopic images, and take thermal and other readings.

Sports Wheelchair An innovative wheelchair

should be designed that is suitable for a variety of sports (e.g. tennis, basketball, rugby). The rear wheels may vary in diameter between 24 and 26 inches. The seat height and back height should be adjustable, so it can be configured for specific sports and disabilities. In addition, the leg rests should be adjustable in length and angle for maximum comfort. Its total weight should be less than 25 pounds. Consideration should also be given to easy transport and storage.

Folding Bicycle A folding bicycle intended

for commuter usage should be designed. The bicycle should have 20-inch wheels and should have a single speed. The weight of the folding bicycle should be less than 20 pounds. The bicycle should approximately fold into a volume measuring $12'' \times 24'' \times 30''$. The cost of the folding bicycle should be under \$500.

Ski-Bike   Ski-biking is a new emerging sport, now permitted at about 50 U.S. resorts. There are currently kits that can be purchased converting any mountain or BMX bike into a high-performance ski-bike. With the increasing popularity of extreme sports, a market for a manufactured ski-bike will become available. The purpose of this project is to develop an affordable (<$1000) ski-bike that can be ridden on the slopes of the U.S. Suitable shock absorbers should be provided. The height of the seat should be adjustable. The weight of the ski-bike should not exceed 30 lbs (remember this has to be carried up a ski lift).

Tour de France Bicycle   A Tour de France

bicycle should be designed for an individual with an approximate height of 180 cm. Its total weight should be less than 10 kilograms but greater than 6.8 kilograms. The wheels may vary in diameter between 55 and 70 cm. Finally, the bicycle shall not measure more than 185 cm in length (including wheels) and 50 cm in width overall.

BMX Bike   The BMX bike should be designed

for race, freestyle/stunt, or dirt/jump. The design will then be largely determined based on its intended use. The frame size and geometry may vary, but the wheels should be 20 inches in diameter and the top tube should be between 20 and 21 inches long.

Dune Buggy   The dune buggy should be capable of negotiating difficult terrain, including both sand and rocks (design is not intended

for street driving). This vehicle is intended for recreation; it should be reasonably safe and stable, and should carry at least two passengers. The dune buggy should be capable of reaching a maximum speed of 60 mph. The fuel tank should have a capacity of at least 5 gallons, with provisions for carrying additional fuel for overnight expeditions. The vehicle should have lights for night driving. Ground clearance will be maximized, with a minimum clearance of at least six inches. Space for stowage of tow ropes, camping gear, and other equipment should be provided. The cost of the dune buggy should not exceed $5000.

All-Terrain Rescue Vehicle   The rescue

vehicle should be able to transport injured individuals, carry medical supplies, and traverse debris. The patient transport area should have enough space to allow the patient to be fully immobilized while two medical officers care for and treat the individual. Several storage containers should be present to store an assortment of equipment (e.g., shovels, chainsaws, and ropes). The vehicle should have mobility to move across extreme land conditions and through a maximum of two feet of standing water.

Fuel-Efficient Family Sedan   The vehicle

should have four doors. It should be as small as possible, but have a minimum length of 180 inches, width of 70 inches, and height of

55 inches. In addition, it should have a minimum wheel base of 100 inches to provide room for comfortable seating. Like all hybrids, it should increase fuel efficiency by harnessing small amounts of electricity generated during braking and coasting.

Backhoe Loader A backhoe should be designed that consists of a tractor, front loader bucket and small bucket in the rear. It should have a dig depth of 15 feet. The loader bucket and backhoe bucket should have capacities of ten and three cubic feet, respectively.

Fire Truck The truck is to be used for

fire-fighting and rescue. The truck should also be equipped with at least a 400-gallon water tank and a 20-gallon foam tank. A ladder lift should allow the rescue of individuals and be able to operate 10 degrees below the horizontal for river rescue or other low-angle situations. In addition, a hose bed, ladder racks, and multiple storage units should be present.

Luxury Train A luxury train is to be designed

to operate at speeds up to 150 mph. The train should consist of six cars with a power car at each end, for a total of eight units. There should be four business-class cars, a first-class car, and a café car. The first-class car should seat approximately 40 passengers and have options for audio and video entertainment. The business-class cars should seat approximately 70 passengers and have an option for audio entertainment only. First class and business class should both include luggage racks and restrooms. The dining car should have the capacity for 34 passengers, and should feature a lounge area and a formal dining area

for meals. In addition, the dining car should have a business center with fax, copy, and print facilities.

Medical Helicopter The helicopter should be

able to rescue individuals from dangerous locations, transport injured individuals, and carry medical supplies. A rescue basket or similar device must be present in order to rescue stranded individuals on water or land. The patient transport area should have enough space for two patients to be fully immobilized while three medical officers care for and treat the individual. Several storage units should be present to store an assortment of equipment (e.g., baskets, ropes, and helmets).

Ultralight Aircraft An ultralight aircraft for

one person should be designed for recreational use. The aircraft should have less than 5 gallons fuel capacity and an empty weight of less than 254 pounds, and the top speed should not exceed 55 knots. Consideration must also be taken for safety.

Commuter Airplane A commuter airplane

capable of transporting a minimum of 35 people should be designed. In addition, the airplane should be designed for two pilots and two flight attendants. Overhead and underseat stowage should be available. The seats should possess adjustable headrests and have the ability to recline. Finally, a lavatory with a baby-changing table should be on board.

Cargo Plane The cargo plane should be able

to transport supplies (e.g., medical, food, and clothing) and equipment, as well as safely drop supplies in remote and/or difficult-to-access areas. The plane should have a payload of at least 20 metric tons, 100 cubic meters capacity, and a range of at least 3,000 kilometers. The cargo transport area should be equipped with adequate means for securing the cargo in place while in transit. A door/hatch that can be easily accessed for dropping the cargo should be present.

Tourism Space Craft A space craft is to be

designed that aims to make a space-flight as routine as a common airline trip. The space craft should be able to travel up to an altitude of 60 miles. In addition to the pilots, it should have the ability to carry at least two passengers. One of the following approaches should be used: (1) taking off horizontally from a runway, (2) lifting off vertically from a platform, (3) being carried or towed in the air by an airplane and launching from the upper atmosphere.

Rescue Submarine The rescue submarine

should have the ability to seek out those in distress. It should be at least 40 feet in length and 12 feet in breadth, in addition to being able to des-

cend to a depth of 3,000 feet. It should be designed to carry a crew of four, with additional space for 20 rescued persons. Several storage racks should be present to store an assortment of equipment (e.g. oxygen tanks, life rafts, and ropes). Finally, the submarine should possess an escape door in case of an emergency.

Sport Fishing Yacht A sport fishing yacht

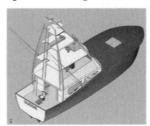

should be designed to be used primarily for billfishing. It should be between 50 and 80 feet. A wide cockpit should be present for the handling of large fish, in addition to a fighting chair. The yacht should be capable of traveling at speeds up to 35 knots. The salon should have a seating area with other amenities (e.g., microwave, fridge, entertainment center).

Fireboat The boat should be designed for a

major port city, and should be at least 100 feet in length. The vessel is to be used for fire-fighting, rescue, and combating oil spills. There should be a minimum of four fire monitors on board, with a combined pumping capacity of at least 15,000 gallons per minute. One of the monitors should be telescoping, in order to combat fires above the waterline. The boat should also be equipped with at least 1,000 gallons of foam concentrate. There should be an emergency medical treatment room, as well as a fast rescue boat with a launching crane. An oil containment boom and other oil spill equipment should also be included.

Hospital Ship A small, fast hospital ship should be designed that will be operated in underdeveloped areas in the Indian Ocean. Under normal operation, the ship will follow a

regular route involving extended stays in different locations. In addition, the ship will have the ability to respond rapidly to disasters anywhere in the Indian Ocean. The hospital will be able to accommodate 50 patients. The ship's design will include a helicopter pad.

Passenger Ferry The passenger ferry vessel for a third-world country experiencing overcrowding should be designed. The ferry should be capable of providing safe transport for up to 100 passengers on routes of no more than three hours' duration. The vessel is intended to provide transportation for the local population.

Roller Coaster The roller coaster has been

popular over decades. Currently Six Flags has the X, a 4-D coaster that provides the standard three degrees of movement with a newly added rotation of the riders, causing a head-over heels effect. Riders are allowed to spin 360 degrees forward and backward independent of the train's primary movement. The purpose of this project is to develop a roller coaster that provides a similar four degrees of freedom. The coaster should provide speeds up to 60 mph, lasting at least 120 seconds. Each train should carry at least 20 people. The track should include at least one 200-foot drop and one 360-degree loop.

Water Theme Park A water theme park is to be developed in a coastal area. The park should consist of several different rides, including slides, pools, an uphill water roller coaster, ring rides, a wave pool, and a flow ride.

Basketball Sports Arena The basketball sports arena/complex should be designed to accom-

modate the men and women's basketball teams. It should be able to seat 15,000 people, of which 1,000 should be high-priority seats. In addition, there should be 25 private suites. Facilities for the spectators should include hospitality rooms, concessions, novelty sales, ticket office, etc. There should be office space to accommodate the athletic administration and building operations. There should be support facilities for both the men and women's basketball teams, including locker rooms, weight and equipment rooms, practice gym, video classrooms, and an academic study center.

Hotel Resort The hotel design should have

a sculpted, organic approach. The resort should have 25–40 standard rooms and 5–10 suites. In addition, the hotel should include such amenities as pools, beach, fitness room, restaurant, marina, tennis courts, etc. The geographic location of the hotel site should be specified.

NASA Launch Pad The launch pad should

consist of a fixed service structure, rotating service structure, access arm, and a mobile launcher platform. The fixed service structure should support the rotating service structure and access arm. The rotating service structure should provide access to the shuttle for the installation and servicing of payloads at the pad. The access arm should allow a crew of six to enter the crew area of the shuttle. Finally, a mobile launcher platform should be designed to bring the shuttle to the pad and serve as the shuttle's launch platform. A remote location for the launch pad should be selected that would provide safety in the event of a disaster.

Modular Space Station The space station,

which is powered by solar energy, should be designed with several modules. Area should be allocated for the following: 10 research laboratories, cargo storage facilities, and living quarters for 20 crew members. A robotic arm (of at least 50 feet in length) should be available for assembly and maintenance tasks on the space station. In the case of an emergency, a crew return vehicle should be present that is capable of holding eight crew members.

Waterway Link The waterway link between the Atlantic Ocean and the North Sea through central Scotland is the host of the $26 million Falkirk Wheel. This is the world's first rotating boatlift and can carry up to four boats, including water. The purpose of this project is to develop a rotating boatlift that will replace a lock system along the Mississippi. It will need to be at least 100 feet in height and be able to support 500 tons of water between the two ends. A 180-degree revolution should be done in less than 20 minutes.

Drawbridge An attractive drawbridge is to be
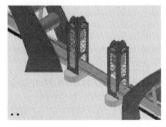
designed at an unspecified site in south Florida, either over an ocean inlet or the Intracoastal Waterway. The bridge roadway is to accommodate two lanes of traffic, with additional lanes provided for cyclists and pedestrians. A bridge tender tower is also to be included in the design. Assume that the span of the waterway is 300 feet, while the horizontal clearance for boat traffic is 90 feet between fenders. There should be a clearance of 25 feet from the underside of the bridge to the water level at mean high tide. Assume an average tidal range of four feet, and that the street level is six feet above mean sea level.

Wind Turbine Energy Farm A wind farm should be designed to provide electricity to a remote area. Site identification in order to maximize the impact is considered to be an important part of the project.

Seaport for a Mid-Size City A port for a mid-size city should be designed for receiving ships and transferring cargo to and from them. An appropriate location along an ocean or sea should be chosen. The port should be equipped with container berths, quay cranes, and warehouse facilities.

Airport for a Mid-Size City An airport for

a mid-size city should be designed to enable airplanes and helicopters to take off and land. The airport should possess runways, taxiways, ramps, and helipads (for helicopters). Other common components should be present, including hangars, terminal buildings, and a control tower. Selection of an appropriate geographic location is a critical portion in the design.

DETAIL DESIGN PROJECTS

Pocket Knife   Design a pocket knife that possesses a variety of blades, scissors, and a screwdriver tip. The weight of the unit should be less than 1.5 ounces.

Sunglasses   Design a pair of sunglasses that not only is lightweight and comfortable, but trendy and unique.

MP3 player   The popularity of digital music and video has reached an all-time high. The purpose of this project is to develop an MP3 player that can play MP3s, but additionally have the ability to display color video (e.g., music videos). The MP3 player should be lightweight (less than six ounces) and should be smaller than a deck of cards.

Calculator   A solar-powered calculator should be designed that possesses basic scientific and trig functions. The calculator should have a 2-line display screen and weigh less than 1.2 pounds.

Travel Mug   A travel mug should be designed to hold a minimum of 16 ounces of coffee. The mug should fit in a standard car cup holder.

Golf Club   Design an iron. Particular attention should be given to the head of the club to create a design that maximizes the *sweet spot*.

Football Helmet   Design a football helmet that would be suitable for the quarterback position.

Remote-Control Toy   A remote-control toy should be designed that has the ability to move forward, backward, left, and right. Potential vehicles for design include: car, truck, boat, or helicopter.

Fishing Rod   A light to medium-action rod and spinning reel should be designed for fresh or salt water fish.

Basketball Hoop   A portable outdoor basketball hoop is to be designed. The height of the hoop should be adjustable, from seven to ten and a half feet. The hoop should be able to withstand 200 pounds of pullover force. In addition, a rim should be designed to allow slamming and dunking without damage (e.g., breaking or bending) to the hoop.

CHAPTER

2 SKETCHING AND OTHER CONCEPTS

□ INTRODUCTION

Prior to the introduction of the personal computer in the early 1980s and the accompanying
introduction of computer-aided design (CAD)
software packages like AutoCAD® shortly
thereafter, most engineering drawings were executed manually using equipment like drafting
machines, T-squares, triangles, compasses, etc.
Today, however, most all engineering drawings
are executed using a CAD system. This sea
change in the way technical drawings are produced has had a major impact on the engineering
graphics curriculum. Instrument drawing (e.g., T-
squares, triangles, etc.), for example, has largely
been replaced by *freehand sketching*.

Engineering is a creative endeavor with roots
that can be traced back to great Italian Renaissance artists like Leonardo, Michelangelo,
Raphaello, and Donatello. While certainly true
that engineering is technology driven, it is important not to lose sight of the discipline's rich
graphical and creative traditions. Great engineering is evidenced as much through the ability to
communicate ideas via freehand sketching, as it is
by manipulating differential equations, making a
computer "sing," or, for that matter, by carefully
reasoned, well-crafted prose.

While some of us are blessed with the natural
ability to draw, most are not. All of us, though, can
improve our ability to communicate, document,
and visualize using freehand sketches. All that is
required is practice.

Freehand sketching has a number of important
uses. Engineers often need to make sketches out
in the field. These sketches are subsequently
converted to CAD back at the office, in order

to document modifications that are to be made to
an existing structure. Sketching is also used as a
just-in-time communication tool, between engineers, technicians, and crafts people, as well as
with clients and supervisors. Freehand sketching
is also used creatively, for brainstorming ideas,
inventing, exploring alternatives, etc. Figure 2-1,

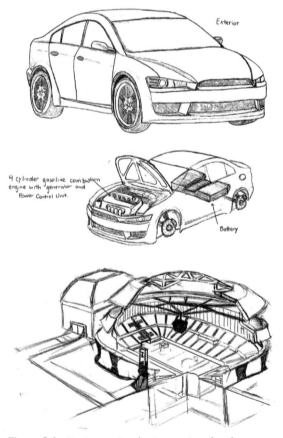

Figure 2-1   Design project brainstorming sketches
(Courtesy of Jonathan Schmid, Donjin Lee)

for example, shows sketches done by engineering graphics students as part of their concept design projects. Either sketching or instrument drawing is essential in order to practice the language of engineering graphics. In her work Sorby[1] has amply demonstrated that freehand sketching is an excellent way to improve spatial visualization skills.

Engineers use technical sketches to create rough, preliminary drawings that represent the main features of a product or structure. Freehand sketches should not be sloppy. Above all, careful attention should be paid to the proportions of the sketch. Although not drawn to a specific scale, freehand sketches should appear to be proportionally accurate to the eye.

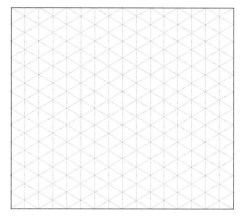

Figure 2-2   Isometric grid paper

☐ SKETCHING TOOLS AND MATERIALS

All that is really needed for sketching is paper, pencils, and an eraser. Almost any kind of paper will do for sketching; the reverse side of already used paper works well. Rectangular grid paper is frequently used for sketching. Engineering firms often use custom-ordered A (8½″ × 11″) or A4 metric size tablets of rectangular grid paper. The grid is actually on the back side of the paper, but is dark enough to be visible on the front side. The paper also includes a custom title block. This paper is useful for combining text, calculations, and sketches.

Isometric grid paper, shown in Figure 2-2, is helpful when first learning to create isometric sketches. This type of grid paper is not typically encountered in engineering firms.

Most engineers use fine-line mechanical pencils, like those shown in Figure 2-3, both for sketching and for calculations. Each mechanical pencil is designed to hold a specific lead size. Common lead sizes include 0.3 mm, 0.5 mm, 0.7 mm, and 0.9 mm. The lead size refers to the diameter of the lead. 0.5 mm is the best size for

Figure 2-3   Fine-line mechanical pencils

general usage, while 0.7 mm is good for bolder strokes. An important advantage of mechanical over wooden pencils is that mechanical pencils do not require sharpening.

Pencil lead is available in different hardness grades, as shown in Table 2-1. The harder the lead, the lighter, sharper and crisper the resulting line. Soft lead, on the other hand, results in darker lines that tend to smear easily. Medium-grade

Table 2-1   Pencil lead grades

Range	Grades	Purpose
Hard	9H, 8H, 7H, 6H, 5H, 4H	Accuracy, precision
Medium	3H, 2H, H, F, **HB**, B	General purpose
Soft	2B, 3B, 4B, 5B, 6B, 7B	Artistic rendering
	harder → softer	

[1] S. Sorby, Developing 3-D Spatial Visualization Skills, *Engineering Design Graphics Journal*, Vol. 63, No. 2, Spring 1999.

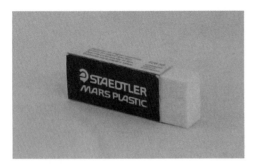

Figure 2-4 Eraser

leads (3H, 2H, H, F, HB, B) are good for general purpose drafting, with HB being the most commonly used lead grade.

Finally, a good eraser that does not tend to smudge is an important sketching tool. See Figure 2-4.

□ SKETCHING TECHNIQUES

Line Techniques

A freehand sketch should begin using proportionally laid out **construction lines**. Construction lines are thin and drawn lightly, and serve to guide the lines to follow. All other lines should be dark, crisp, and of uniform thickness, and can be sketched directly over the construction lines. See Figure 2-5, where construction lines are first used to lay out the proportions of the sketch, and then the features are indicated using bold lines. If drawn correctly, construction lines need not be erased. It is a common mistake to make construc-

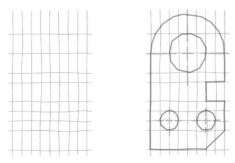

Figure 2-5 Construction lines with bold lines drawn over

tion lines that are too dark, making them hard to distinguish from other lines.

In addition to the relative lightness or darkness of a line, two line widths, thick and thin, are used in engineering sketches and drawings. Continuous lines indicating visible object edges are thick, while hidden, center, and construction lines are thin. See Figure 2-6 for a brief summary of the characteristics of the most commonly used lines.

Figure 2-6 Common sketching line types

A more thorough discussion of the line styles employed in CAD and manual engineering drawings is provided at the end of the chapter.

Sketching Straight Lines

To sketch a straight line, start by marking both end points. Next, place the pencil point at one of these end points. Keeping your eye fixed on the point to which the line is being drawn, sketch a light line with a single stroke. Finally, darken the line. See Figure 2-7.

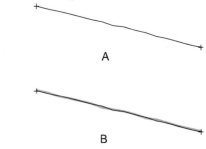

Figure 2-7 Sketching a straight line

To sketch an especially long line, use several short overlapping strokes, and then darken the line, as seen in Figure 2-8. Slight wiggles are okay as long as the resulting line is straight. Occasional

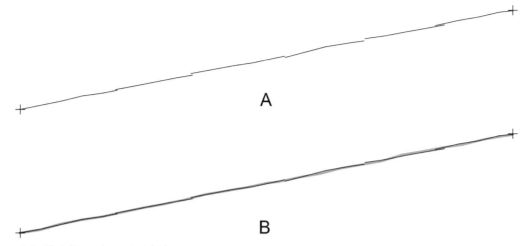

Figure 2-8   Sketching a long straight line

gaps are also acceptable. See Figure 2-9. If the resulting line tails off and is not straight, the pencil is probably being gripped too tightly. See Figure 2-10 for an example.

Figure 2-9   Wiggles and gaps in sketched lines

Figure 2-10   Sketched line tailing off due to overly firm grip

For right-handers, horizontal line strokes are typically made from left to right, as shown in Figure 2-11. Vertical lines are usually drawn from top to bottom, as seen in Figure 2-12. Inclined lines at certain orientations can be difficult to draw, in which case the paper can be rotated to a more comfortable position and then drawn, as shown in Figure 2-13. In fact,

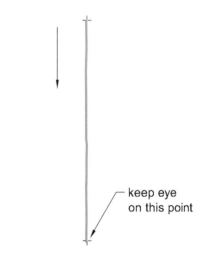

Figure 2-12   Sketching vertical lines

keep eye on
end point

Figure 2-11   Sketching horizontal lines

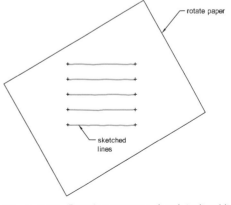

Figure 2-13   Rotating paper to sketch inclined lines

for lines drawn at any orientation, it is a good idea to rotate the paper to a comfortable position, and then sketch the line.

Sketching Circles

A number of different techniques can be used to sketch a circle. Using the trammel method, a trammel, or straight-edged piece of scrap paper, is used to locate points on the circumference of the circle. On a straight edge of the paper, mark two points at a distance equal to the radius of the circle. With one point at the center, rotate the paper and mark as many points on the circumference as desired; then sketch a circle passing through the points. Figure 2-14 shows the trammel method.

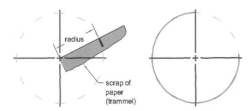

Figure 2-14   Trammel method used to sketch circles

Using the square method, the enclosing square of the circle is first sketched. Next the midpoints of the sides of the square are marked. The midpoints are used as the quadrants of the circle. A circle passing through the quadrant points and tangent to the sides of the square is then sketched. See Figure 2-15 for an example of a circle constructed using the square method. Note that by constructing the diagonals of the square and marking the radius along the diagonals, additional points on the circumference of the circle can be added.

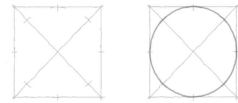

Figure 2-15   Square method used to sketch circles

Either of these methods can be used to sketch a circle or arc feature in a *pictorial sketch* (see

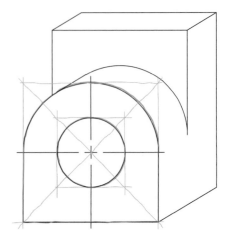

Figure 2-16   Circles and arcs sketched on an oblique pictorial

Chapter 3), as long as the face on which the circle or arc appears is parallel to the view plane. Figure 2-16 shows an example of this for an *oblique* pictorial sketch.

Sketching Ellipses

The rectangle method can be used to sketch an ellipse. First construct the enclosing rectangle, and then locate the midpoints along the sides of the rectangle. Sketch the ellipse passing through the midpoints and tangent to the sides of bounding rectangle. See Figure 2-17 for an example of this method.

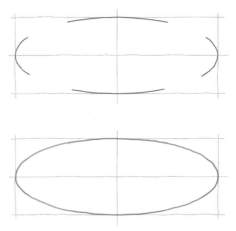

Figure 2-17   Rectangular method used to sketch ellipses

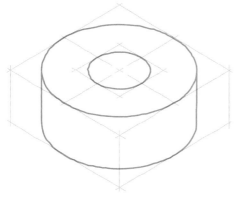

Figure 2-18   Parallelogram method used to sketch ellipses

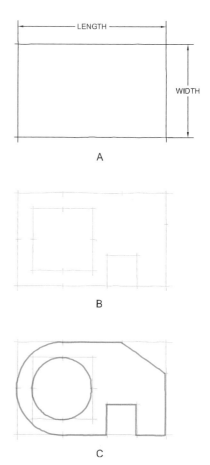

A

B

C

Figure 2-19   Proportioning with principal dimensions

In a pictorial sketch a circle will appear as an ellipse, unless the circle is parallel to the view plane. To sketch such an ellipse, first construct the enclosing parallelogram. The ellipse will pass through the midpoints of the sides of the parallelogram, and be tangent to the sides of the parallelogram. Figure 2-18 shows an example of the construction of an ellipse on an *isometric* pictorial sketch.

□ PROPORTIONING

Although freehand sketches are not drawn to scale, it is important to maintain the relative proportions between the principal dimensions of the object. To accomplish this, first estimate the proportional relationship between the object's principal dimensions and lightly block them in, as shown in Figure 2-19A. The estimated dimensions of each feature should then be proportioned with respect to established dimensions, as seen in Figure 2-19B. Work at developing the ability to divide a line in half by eye; the halves can then be further divided into fourths. Finally, use bold lines to fully define the object, as seen in Figure 2-19C.

While it is assumed that scales are not used when sketching, a *trammel* may be used to improve proportional accuracy. Decide upon a unit of length, and then transfer it to the trammel. Additional unit-length graduations can be added

to the trammel, which can then be used as a scale to lay out the proportions of each object feature. See Figure 2-20 on the following page for an example of this technique. Note also that the trammel can be folded to obtain half and quarter lengths.

Estimating Dimensions of Actual Objects

It is sometimes necessary to make a sketch of an actual object. To make such a sketch, hold the pencil at arm's length, with the pencil between the eye and the object. Using the pencil as a sight, establish a proportional relationship between an object edge and the length of the pencil. Do this by aligning the end of the pencil with one end of

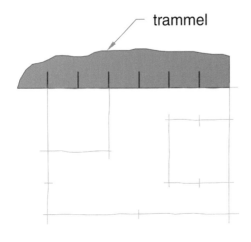

trammel

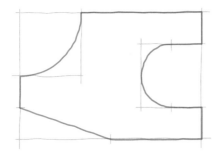

Figure 2-20 Using a trammel to improve proportions

the object edge. Move your thumb along the pencil until it coincides with the other end of the edge. Now use this proportion to estimate the lengths of other object features. See Figure 2-21 for an example of this technique. A similar tech-

Figure 2-21 Using a pencil to estimate proportional lengths

Figure 2-22 Using a pencil to estimate angles

nique can be used to estimate angles, as shown in Figure 2-22.

Partitioning Lines

The following procedure may be used to divide a line into fractional parts. To subdivide the line AB shown in Figure 2-23, construct a rectangle on the line AB. Next draw both of the diagonals of the rectangle. Pass a line perpendicular to AB that passes through the intersection of the diagonals. This line divides AB in half.

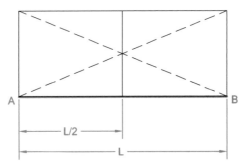

Figure 2-23 Method for partitioning lines

To partition AB into thirds, sketch another diagonal from one corner of the original rectangle to the midline on the opposite side. Now sketch a line perpendicular to AB that passes through the point where the two diagonals (full and half diagonal) intersect. This determines a one-third length of AB, as seen in Figure 2-24. Repeating this same process will further divide the line into fourths, fifths, etc. Figure 2-25 shows this.

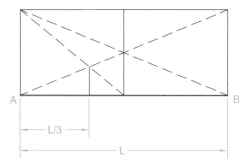

Figure 2-24 Method for partitioning lines into thirds

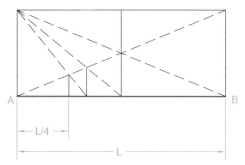

Figure 2-25 Method for partitioning lines into quarters

□ INSTRUMENT USAGE— TRIANGLES

Although drafting machines and T-squares have fallen into disuse, the construction of parallel and perpendicular lines using triangles remains a useful skill. Every engineer should possess both a 45° and a 30° – 60° triangle. These triangles are shown in Figure 2-26.

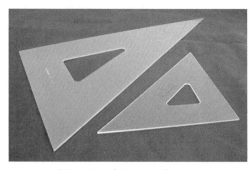

Figure 2-26 30°– 60° and 45° triangles

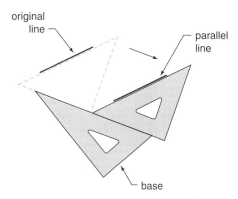

Figure 2-27 Using triangles to draw parallel lines

Parallel Lines

In order to draw a line parallel to another line using two triangles, align the hypotenuse of one triangle with the original line. Next place the hypotenuse of the second triangle along the leg of the first triangle. Now slide the first triangle along the fixed second triangle until the location of the parallel line is reached, and construct the line. Figure 2-27 demonstrates this technique.

Perpendicular Lines

Triangles are also useful for drawing one line perpendicular to another. There are a couple of ways to accomplish this. In the first method, shown in Figure 2-28, one leg of a triangle is

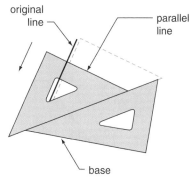

Figure 2-28 Using triangles to draw perpendicular lines: method 1

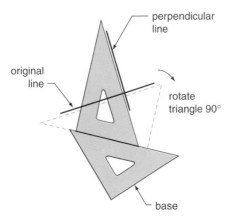

Figure 2-29 Using triangles to draw perpendicular lines: method 2

aligned with the original line, while a second triangle is used as a base. By sliding the first triangle along the base triangle, the other leg of the first triangle can be used to draw the perpendicular line. In the second method, shown in Figure 2-29, the first triangle is rotated 90°, rather than slid, prior to drawing the perpendicular line.

☐ INSTRUMENT USAGE—SCALES

Introduction

In the language of engineering graphics, the word *scale* is used in a variety of different ways. One meaning of the verb scale is to transform an object by making it proportionally larger or smaller. By using, for example, the AutoCAD® SCALE command, an object can be made proportionally larger or smaller.

As a noun, a scale refers to a proportion between two sets of dimensions, as for example, between those of a drawing of an object and the actual object. Figure 2-30 shows an example of

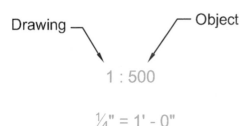

Figure 2-30 Drawing scale examples

two such scales. While in metric units a colon is typically used to separate the two sets of dimensions, in English units an equal (=) sign is used.

On graphs and charts, the term scale refers to a sequence of marks laid down at predetermined distances along an axis. The marks are called **graduations**. The numerical values assigned to significant graduations are called **calibrations**. See Figure 2-31. Note also that there are both major and minor graduations.

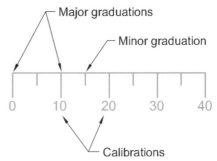

Figure 2-31 Calibrations, graduations, majors and minors

Finally, a scale is an instrument with one or more sets of spaces graduated and numbered on its surface for measuring or marking off distances or dimensions. Scales may have either a flat or a triangular cross section, as shown in Figure 2-32.

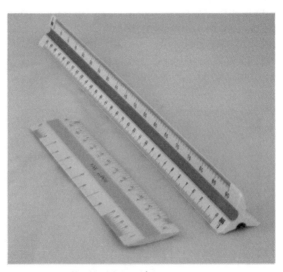

Figure 2-32 Engineering scales

There are several different kinds of scales, classified both by discipline (e.g., mechanical, architectural, civil), and by the system of units employed (English or metric). All scales are designed so that, given the drawing scale, dimensions can be read directly from the scale, without resorting to calculation.

Note that although similar in appearance to a ruler, a scale should only be used for measurement purposes, and not as a straight edge. Improper use of a scale can damage the precisely etched graduations on the face of the scale.

Engineer's Scale

The engineer's scale is used to measure large structures, like dams and bridges, commonly worked on by civil engineers. The engineer's scale is a decimal scale. On a triangular engineer's scale, a measured inch is divided into 10, 20, 30, 40, 50, or 60 parts. See Figure 2-33. Each face on the scale can be used for multiple scales. For

example, the 20 scale (where an inch is divided into 20 parts) can be used for any of the following scales: $1'' = 20'$, $1'' = 2'$, $1'' = 200'$, $1'' = 2$ miles, etc. Since there is typically a mismatch between the drawing and the real-world units when using this scale (e.g., $1'' = 20'$ means an inch on the drawing is equivalent to 20 feet in the real world), the *scale factor*, or the amount that the object is scaled up or down to fit it on a sheet of paper, is not 20 but rather $20*12 = 240$. In other words, with a scale of $1'' = 20'$, the object has been scaled down to 240 times its original size in order to fit it on a sheet of paper.

Engineer's scale example

Assume that a line is drawn to the scale $1'' = 20'$. Determine the actual length that the line represents.

Using the 20 scale, we know that one inch on the drawing represents 20 feet in the real world. Further, each minor graduation represents 1 foot (i.e., 20 feet/ 20). Laying the 20 scale along the line as shown in

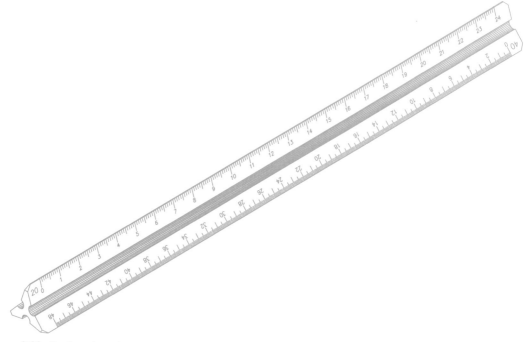

Figure 2-33 Engineer's scale

Figure 2-34 Engineer's scale example

Figure 2-34, we find that the line represents an actual length of 81 feet.

Architect's Scale

As the name implies, an architect's scale is used to scale drawings of buildings like houses, schools, libraries, etc. Some common architect's scales include $\frac{1}{4}'' = 1' - 0''$, $\frac{1}{2}'' = 1' - 0''$, and $1'' = 1' - 0''$. The scale $\frac{1}{4}'' = 1'' - 0''$ means that a $\frac{1}{4}''$ on the drawing represents one foot in reality. An object drawn at this scale is consequently scaled down 48 times ($\frac{1}{4}'' = 12''$) in order to fit it on a drawing sheet.

An architect's scale is divided into two sections: a larger section for measuring feet, and a smaller subsection for measuring inches. See Figure 2-35. Two scales, for example 1/8 and 1/4 to the foot, share the same face on a scale. One scale is read from left to right, the other from right to left.

Architect's scale example

Assume that a line is drawn to the scale $\frac{1}{4}'' = 1' - 0''$. Determine the actual length that the line represents.

Using the $\frac{1}{4}$ scale, align the scale so that one end of the line falls within the subsection area of the scale, while the other end of the line is aligned with a foot calibration on the scale, as shown in

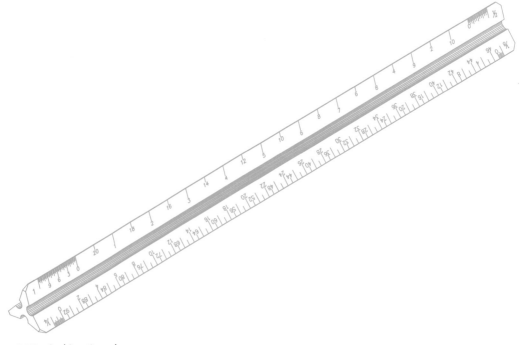

Figure 2-35 Architect's scale

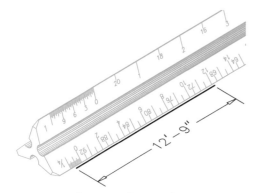

Figure 2-36   Architect's scale example

Figure 2-36. This line represents a real-world length of 12 feet, 9 inches.

Mechanical Scale

A mechanical scale is used to scale smaller parts commonly designed by mechanical engineers. A mechanical scale is similar in appearance to an architect's scale; it is also divided into a larger section and a subsection. On a mechanical scale, however, the large section measures inches, while the smaller subsection measures fractions of an inch. See Figure 2-37. Common mechanical scales include $1'' = 1''$ (full size), $\frac{1}{2}'' = 1''$ (half size), $3/8'' = 1''$, and $\frac{1}{4}'' = 1''$. Note that the word SIZE appears on mechanical scales, and can be used to help distinguish them from architect's scales. Size in this context has the same meaning as scale factor; that is, $\frac{1}{4}$ SIZE literally means that the drawing of the object is scaled to one fourth the size of the actual object.

Mechanical scale example

Assume that a line is drawn to the scale $3/8'' = 1''$. Determine the actual length that the line represents.

Using the 3/8 SIZE scale, align the scale so that one end of the line falls within the subsection area of the scale, while the other end of the line is aligned with an inch calibration on the scale, as shown in

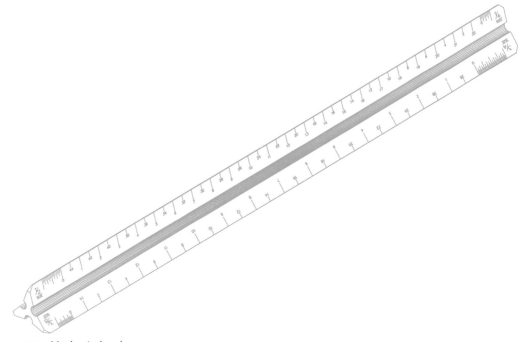

Figure 2-37   Mechanical scale

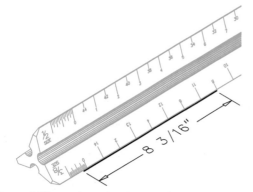

Figure 2-38 Mechanical scale example

Figure 2-38. This line represents a real-world length of 8-3/16 inches.

Metric Scale

The three previously discussed scales, engineer's, architect's, and mechanical, all use English units. There are consequently several different triangular metric-scale designs, intended especially to meet the needs of architects, civil, and mechanical engineers. A commonly used metric triangular scale intended mainly for scaling large structures has the following scales: 1:100, 1:200, 1:250, 1:300, 1:400 and 1:500. See Figure 2-39. Note that the units of the calibrations on this scale are in meters.

Metric scale example 1

Assume that a line is drawn to the scale 1:300. Determine the actual length that the line represents.

Using the 1:300 face, align the scale with the line, as shown in Figure 2-40. The line represents an actual length of 26.6 meters (or 26,600 mm).

Metric scales are most similar to an engineer's scale. Like an engineer's scale, one face on the scale, say 1:200, can be used for multiple scales (e.g., 1:2, 1:20, or 1 mm = 2 m).

Metric scale example 2

Assume that a line is drawn to the scale 1:30. Determine the actual length that the line represents.

Since the calibrations when reading a 1:300 scale are in meters, using the same face as a 1:30 scale

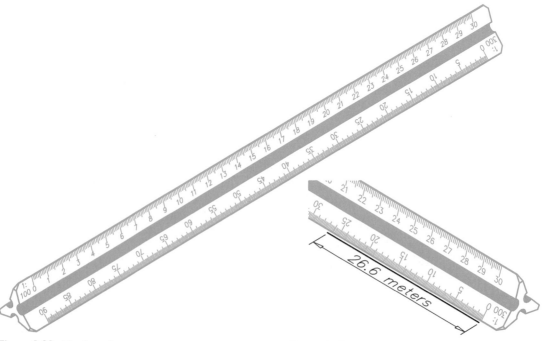

Figure 2-39 Metric scale

Figure 2-40 Metric scale example 1

means that the calibrations are in decimeters (i.e., mm × 100). Aligning the scale with the line, as shown in Figure 2-41, we see that the line represents an actual length of 3340 millimeters (3.34 meters).

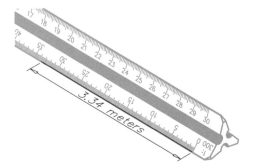

Figure 2-41   Metric scale example 2

□ LINE STYLES

Line styles, sometimes referred to as the alphabet of lines, describe the size, spacing, construction and application of the various lines used in CAD and manual engineering drawings. These conventions are established in ASME Y14.2M-1992, Line Conventions and Lettering.

Two line widths are recommended on engineering drawings, thick and thin. The ratio of these line widths should be approximately 2:1.

The thin line width is recommended to be a minimum of 0.3 mm, with a recommended minimum thick line width of 0.6 mm.

Figure 2-42 shows the various types of lines, and their widths. *Visible lines* are used to show the visible edges and contours of objects. *Hidden lines* are used to show the hidden edges and contours of objects. *Center lines* are used to represent axes or center planes of symmetrical parts and features, as well as bolt circles and paths of motion. *Phantom lines* are most commonly used to show alternate positions of moving parts. *Cutting-plane lines* are use to indicate the location of cutting planes for sectional views. The ends of the lines are at 90 degrees, and are terminated by arrowheads to indicate the viewing direction. *Section lining* is used to indicate the cut surfaces of an object in a section view. *Break lines* are used when complete views are not needed.

Dimension lines are used to indicate the extent and direction of a dimension, and are terminated with uniform arrowheads. *Extension lines*, used in combination with dimension lines, indicate the point or line on the drawing to which the dimension applies. *Leader lines* are used to direct notes, dimensions, symbols, and part numbers on a drawing. A leader line is a straight inclined line, not vertical or horizontal, except for a short horizontal portion extending to the note.

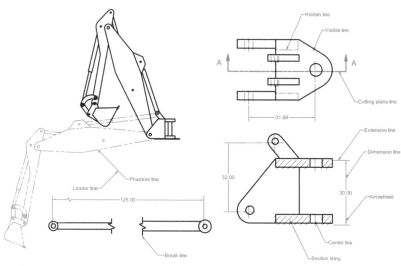

Figure 2-42   Line styles

Arrowheads are used to terminate dimension, leader and cutting plane lines. Arrowhead length and width should be a ratio of approximately 3:1. A single style of arrowhead should be used throughout the drawing.

☐ QUESTIONS

TRUE AND FALSE

1. If an architect's scale with a 1/4 identifier is used to measure a drawing of an object, then the scale factor will be 1/4 (i.e., scaled down 4 times).

2. Using a scale with the identifier 20, the distance between the calibrations 2 and 8 could represent an actual length of 6, 60, or 600 feet.

SKETCHING

3. Using the cues provided, sketch straight lines for:
   a. Figure P2-1   b. Figure P2-2

4. Using the cues provided, sketch the nested shapes for:
   a. Figure P2-3   b. Figure P2-4

5. Using a trammel, sketch the missing symmetrical half of the object for:

   a. Figure P2-5   b. Figure P2-6
   c. Figure P2-7   d. Figure P2-8
   e. Figure P2-9   f. Figure P2-10

6. Using a trammel and the cue provided, sketch a replica of the object for:
   a. Figure P2-11   b. Figure P2-12

SCALES

7. Identify the type of scale shown in:
   a. Figure P2-13   b. Figure P2-14
   c. Figure P2-15   d. Figure P2-16

8. Determine the actual lengths from Figure P2-17 using a scale of:
   a. $1/4'' = 1''$ b. $1/2'' = 1''$ c. 1:5   d. 1:10

9. Determine the actual lengths from Figure P2-18 using a scale of:
   a. $3/8'' = 1''$ b. $3/4'' = 1''$ c. 1:3   d. 1:400

10. Determine the actual lengths from Figure P2-19 using a scale of:
    a. $1'' = 400'$ b. $1'' = 30'$ c. $1/2'' = 1' - 0''$
    d. $1/8'' = 1' - 0''$

11. Determine the actual lengths from Figure P2-20 using a scale of:
    a. $1'' = 500'$ b. $1'' = 6'$ c. $1/4'' = 1' - 0''$
    d. $1'' = 1' - 0''$

Note: Problems P2-1 through P2-12 can be downloaded from www.wiley.com/college/leake

Figure P2-1

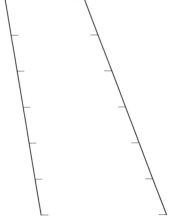

Figure P2-2

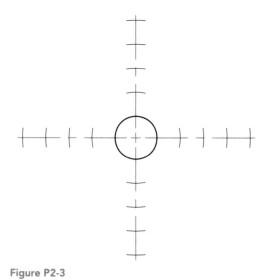

Figure P2-3

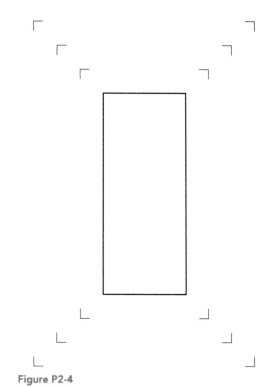

Figure P2-4

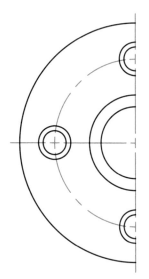

Figure P2-5

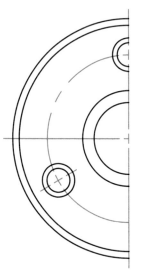

Figure P2-6

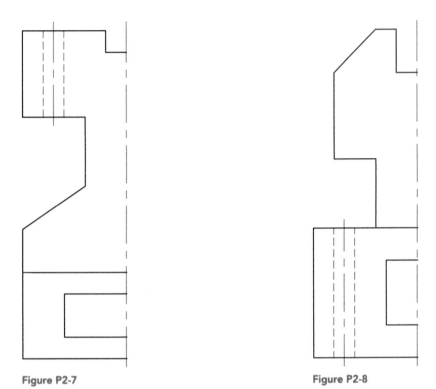

Figure P2-7

Figure P2-8

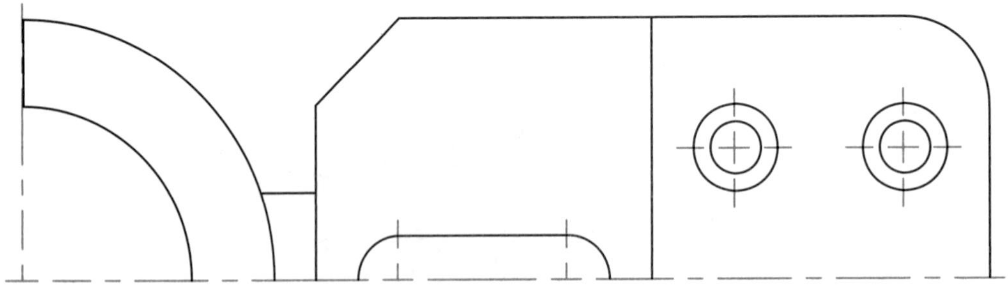

Figure P2-9

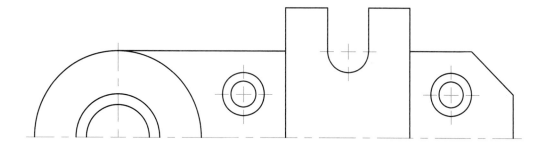

Figure P2-10

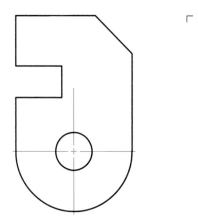

Figure P2-11

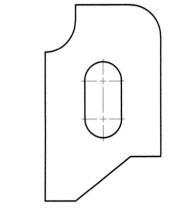

Figure P2-12

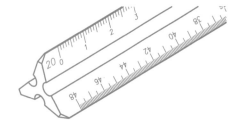

Figure P2-13

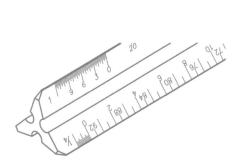

Figure P2-14

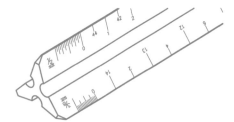

Figure P2-15

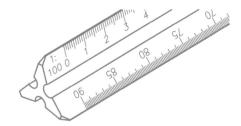

Figure P2-16

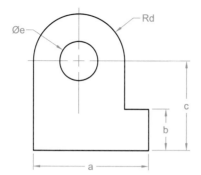

Figure P2-17

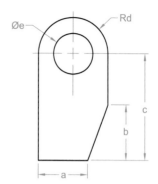

Figure P2-18

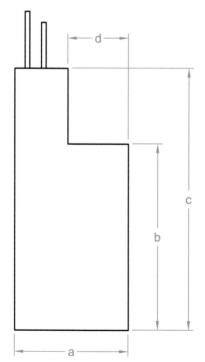

Figure P2-19

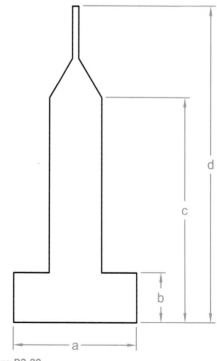

Figure P2-20

3 PLANAR PROJECTIONS AND PICTORIAL VIEWS

☐ PLANAR PROJECTIONS

Introduction

Projection is the process of reproducing a spatial object on a plane, curved surface or line by projecting its points. Common examples of projection include photography, where a 3D scene is projected onto a 2D medium, and map projection, where the earth is projected onto either a cylinder, a cone, or a plane in order to create a map. *Planar projection* figures prominently in both engineering and computer graphics. For our purposes, a projection is a mapping of a three-dimensional (3D) space onto a two-dimensional subspace (i.e., a plane). The word projection also refers to the two-dimensional (2D) image resulting from such a mapping.

Every planar projection includes the following elements:

- The 3D object (or scene) to be projected
- Sight lines (called projectors) passing through each point on the object
- A 2D projection plane[1]
- The projected 2D image that is formed on the projection plane

These elements are indicated in Figure 3-1. The projection is formed by plotting piercing points created by the intersection of the projectors with the projection plane. By mapping these points onto the projection plane, the 2D image is formed. In effect, three-dimensional information is collapsed onto a plane.

[1] Although commonly represented as a bounded rectangle, the projection plane is infinite in extent.

The Albrecht Dürer drawing seen in Figure 3-2 on page 51 further illustrates these projection elements. The lute lying on the table is the *object*, the piece of string is a (movable) *projector*, and the frame through which the artist looks is the *projection plane*. Two movable threads are mounted on the frame, allowing the artist to identify a single piercing point. A piece of paper hinged to the frame serves as the basis for the *projected image*. Once the projected point is transferred to the paper, the artist's assistant moves the string to another point on the lute and the process is repeated.

Classification of Planar Projections: Projector Characteristics

Planar projections are initially classed according to the characteristics of their projectors. In a *perspective projection*, the projectors converge to a single viewpoint called the center of projection (CP). The *center of projection* represents the position of the observer of the scene, and is positioned at a finite distance from the object. Dürer's *Artist Drawing a Lute* (Figure 3-2) illustrates

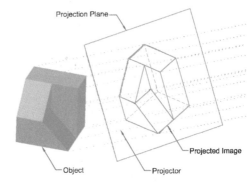

Figure 3-1 Elements of a planar projection

Figure 3-2 Albrecht Durer, *Artist Drawing a Lute*, 1525

perspective projection, with an eyebolt mounted on the wall serving as the center of projection. If the center of projection is infinitely far from the object, the projectors will be parallel to one another. In this case a ***parallel projection*** results. Figure 3-3 compares the projectors in both perspective and a parallel projection, where in this case the projected object is simply a vertical line.

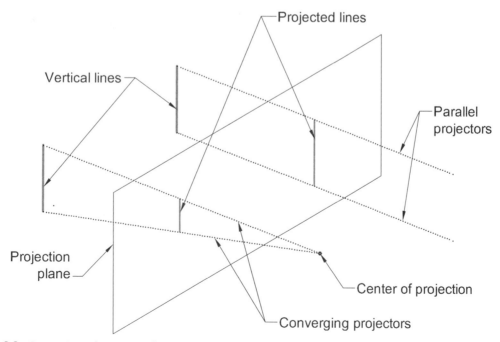

Figure 3-3 Comparison of projectors for perspective and parallel projection

Preliminary Definitions

Before moving on to a more detailed discussion of the different kinds of planar projections, it is useful to introduce some additional terminology. Figure 3-4 shows a cut block, typical of the objects used throughout this work. An object's *principal*

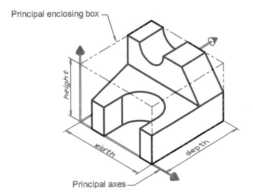

Figure 3-4 PEB, principal dimensions, and principal axes

enclosing box (PEB) just contains the object, and consequently its dimensions are the maximum width, depth, and height of the object. These are referred to as the *principal dimensions* of the object. Mutually perpendicular axes corresponding to the edges of the PEB are referred to as the *principal axes* of the object. The PEB is also referred to as a *bounding box*; both terms can be used interchangeably.

The faces of the PEB are referred to as the *principal planes* or *faces* of the object. As shown in Figure 3-5, these planes are categorized as being either frontal (front, back), horizontal

(top, bottom), or profile (right, left), for a total of six.

Foreshortening is an important graphical concept related to projection theory. A dictionary definition of *foreshorten* is to shorten by proportionately contracting in the direction of depth so that an illusion of projection or extension in space is obtained. To demonstrate foreshortening, close one eye and hold your hand in front of you, with your palm perpendicular to your line of sight. Now curl your fingers toward you until they are parallel to your line of sight. The image that you see of your fingers is foreshortened.

The term *pictorial* is used to indicate a kind of projection, view, drawing, or sketch that includes all three dimensions, and in consequence provides the illusion of depth. The principal types of pictorial views are perspective, oblique, and axonometric. Figure 3-6 on page 53 shows four different pictorial views of the same object. The two on the left (i.e., A. Oblique, C. Isometric) are parallel projections, while the two on the right (i.e., B. One-point, D. Two-point) are perspective projections.

Block coefficient

In ship design, the ratio of a ship's displaced volume to that of its length, width, and depth up to the waterline (i.e., the PEB) is called the *block coefficient* (see Figure 3-7 on page 53). The block coefficient can be used to compare the relative fineness between different hull forms. For example, a speedboat or destroyer might have a block coefficient of about 0.38, while an oil tanker has a block coefficient of about 0.80.

Classification of Planar Projections: Orientation of Object with Respect to Projection Plane

Beyond the characteristics of their projectors, planar geometric projections can be further categorized, as shown in Figure 3-8 on page 54 . As we will see, these planar projection subclasses are in large part based on the orientation of the object with respect to the projection plane.

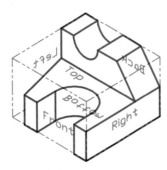

Figure 3-5 Principal planes

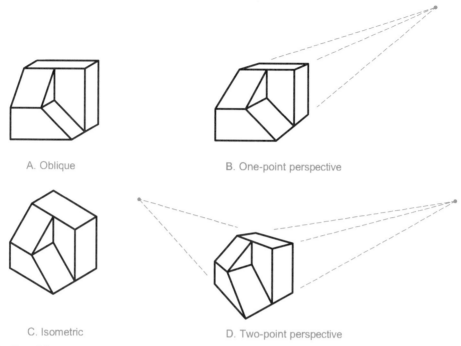

A. Oblique

B. One-point perspective

C. Isometric

D. Two-point perspective

Figure 3-6 Pictorial projections

In Figure 3-9 on page 54, three bounding boxes are shown in different orientations with respect to a vertical projection plane. Figure 3-10 on page 54 shows these same elements when viewed from above. As a result the projection plane is now seen on edge. Note that the principal axes of each bounding box are also shown. For box A on the left, two principal axes are parallel to the projection plane, while the third axis is perpendicular to the projection plane. In the case of box B in the center, one principal axis (i.e., vertical), is parallel to the projection plane, while the other two are inclined (i.e., neither parallel nor perpendicular) to the projection plane. In the last case, box C on the right, all three axes are inclined to the projection plane. These possible orientations, together with the type of projection, either perspective or parallel, determine most of the planar projection categories. Figure 3-11 on page 55 is based on Figure 3-8 but includes images depicting the orientation of the object with respect to the projection plane.

Further Distinctions Between Parallel and Perspective Projections

In a parallel projection, the center of projection is infinitely far from the object being projected. This means that the projectors are now parallel to one another. While perspective projection is useful in creating a more realistic depiction of an object or scene, a parallel projection is typically used when it is important to preserve the dimensional properties of the object. As an example, compare the projections employing parallel projectors in Figure 3-12 on page 56 with the converging projectors used in Figure 3-13 on page 56. In both

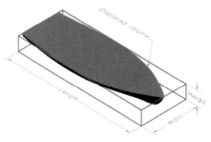

Figure 3-7 Block coefficient

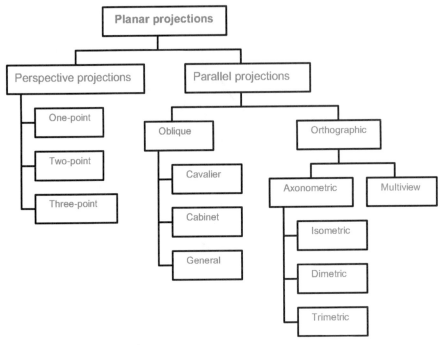

Figure 3-8   Planar geometric projection classes

cases the projected object face is parallel to the projection plane. Notice that parallel projection preserves both the size and shape of the object face, while perspective projection only preserves the shape of the object face (and the shape may even be inverted!), depending upon the location of the projection plane.

In a parallel projection, parallel object edges remain parallel when projected. Unlike perspective projection, there are no **vanishing points**. Figure 3-14 on page 57 shows a parallel with a perspective projection of the same object. Notice that while the receding edges of the parallel projection on the left remain parallel, the corresponding edges on the perspective projection on the right converge to a vanishing point. While the perspective projection provides a more realistic depiction of the object, the parallel projection, by preserving parallelism, is easier to scale.

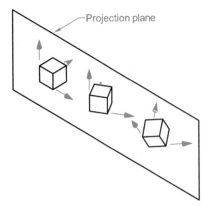

Figure 3-9   Orientation of object with respect to projection plane

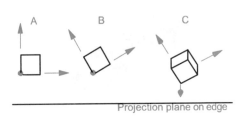

Figure 3-10   Orientation of object with respect to projection plane as seen from above

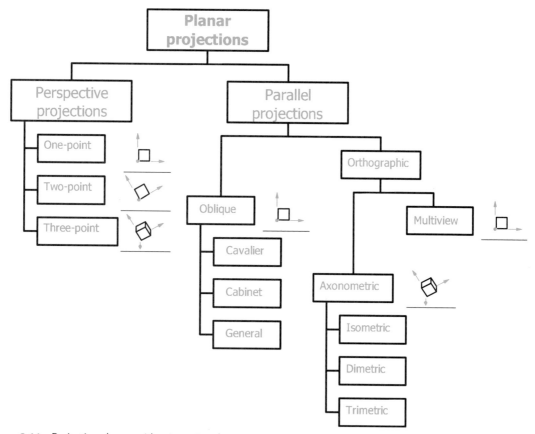

Figure 3-11   Projection classes with orientation shown

While extremely important in computer graphics, perspective projection is not commonly employed in engineering. For this reason perspective projection and perspective sketching are treated separately in Appendix A.

Classes of Parallel Projections

Figure 3-15 on page 57 shows a breakdown of the different kinds of parallel projection techniques. Oblique and axonometric projections will be discussed in the remainder of this chapter. Multiview sketching is the subject of the following chapter.

☐ OBLIQUE PROJECTIONS

Oblique drawings are traditionally employed when one object face is significantly more complicated than the other faces of the object.

Oblique Projection Geometry

Figure 3-16 on page 58 shows the geometric arrangement of an oblique projection. As seen in Figure 3-16A, parallel projectors intersect the projection plane at an oblique angle. In addition, one principal face of the object is parallel to the projection plane. In an oblique projection, the object face that is parallel to the projection plane is projected true size.

Oblique Projection Angle

Two angles can be used to describe the intersection of an oblique projector with a projection plane, as shown in Figure 3-17 on page 58. An in-plane angle β measures the angle of rotation of the projector about the projection plane normal. The out-of-plane angle α is called the *oblique projection angle*.

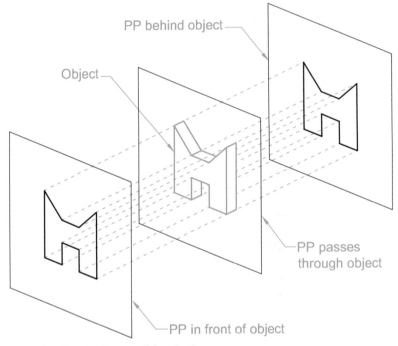

Figure 3-12 Projection plane location in a parallel projection

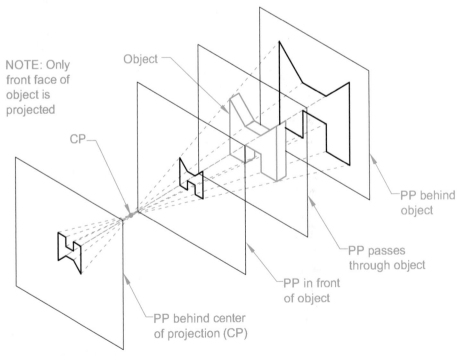

Figure 3-13 Projection plane location in a perspective projection

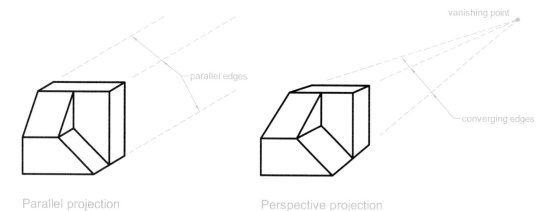

Figure 3-14 **Comparison of parallel and perspective projections**

The oblique projection angle determines the type of oblique projection: cavalier, cabinet, or general.

Classes of Oblique Projections

Figure 3-18 on page 59 shows an oblique projection setup, where two identical cubes (shown in the top half of the figure) are projected onto a projection plane (in the bottom half of the figure). The front face of both cubes is projected true size.

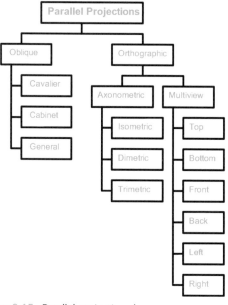

Figure 3-15 **Parallel projection classes**

The oblique projection angle (α) used for the cube on the left is 45 degrees, and the resulting projection is called a ***cavalier oblique***. Notice that the projection of this cube appears to be elongated in the receding (depth) axis direction. In fact, the measure of all of the edges of both the cube and its cavalier projection is identical. In a cavalier projection the receding axis is not foreshortened; it is scaled the same as the other (horizontal, vertical) principal axes.

On the right half of Figure 3-18, the oblique projection angle (α) is approximately 63.43 degrees, resulting in a ***cabinet oblique*** projection. In comparing the projected lengths of this cube, you will find that the projected receding edge length is one half that of the other (horizontal, vertical) edge lengths. The receding axis of a cabinet oblique is foreshortened to exactly one half that of the other principal axes. Notice also that the projection that results from a cabinet oblique appears to the eye to be more proportionally correct than the cavalier oblique. This is because visually we expect some foreshortening to occur along a receding axis.

If an oblique projection angle between 45 and 63.43 degrees is used, the result is called a ***general oblique*** projection. On a general oblique, the receding axis is scaled between ½ and 1. This would normally be done to improve the appearance of the projected image. Table 3-1 summarizes the different kinds of oblique projections.

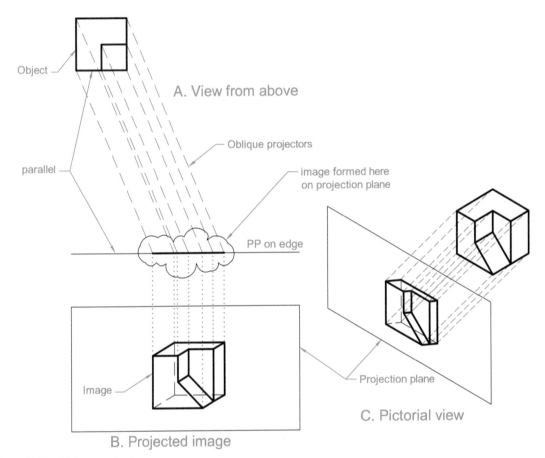

A. View from above

Object

Oblique projectors

parallel

image formed here
on projection plane

PP on edge

Projection plane

C. Pictorial view

Image

B. Projected image

Figure 3-16   Oblique projection geometry

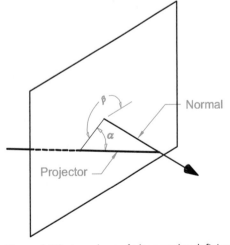

Normal

β

α

Projector

Figure 3-17   In and out of plane angles defining an
oblique projection

Table 3-1   **Classes of oblique projections**

Type of Oblique	Oblique Projection Angle (α)	Receding Axis Scale
Cavalier	45°	1
Cabinet	63.43°	½
General	45° < angle < 63.43°	½ < scale < 1

Oblique projection angle in 2D

By collapsing the 3D oblique projection geometry
into 2D, it is easy to understand why a 45-degree
projection angle results in no foreshortening (i.e., 1:1
scale), while a ~63.43 degree angle scales a projected
length to one half of the original. Figure 3-19 on page

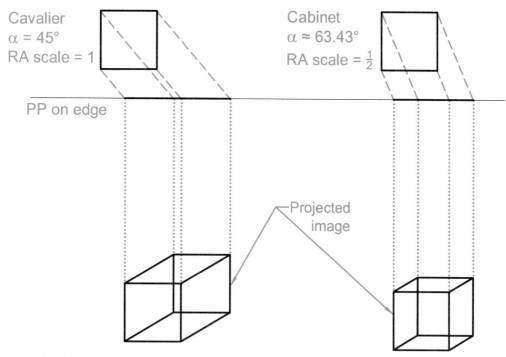

Figure 3-18   Cavalier versus cabinet oblique

59 shows a projection plane and the object, a line of length L. The line is perpendicular to the projection plane, much as the cube edges (not parallel to the projection plane) in Figure 3-18 are perpendicular to the projection plane.

Note: Tan 45° = 1   Cot 63.43° = ½

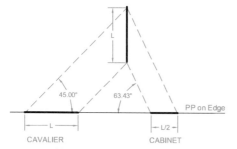

Figure 3-19   Oblique projection angle in 2D

Receding Axis Angle

In an oblique drawing, one axis is horizontal and another is vertical. The third receding (or depth) axis can be inclined at any angle, but is normally chosen to be 30, 45 or 60 degrees. This angle determines the relative emphasis of the receding planes on the projection, as seen in Figure 3-20 on page 60. Note that the receding axis angle should not be confused with the oblique projection angle.

The receding axis angle is related to the in-plane projector angle β discussed earlier and shown in Figure 3-17. As seen in Figure 3-21 on page 60, the receding axis angle is equal to $\beta - 180$ degrees, where β is the in-plane angle of rotation of the oblique projector about the projection plane normal.

□ ORTHOGRAPHIC PROJECTIONS

Orthographic projection is the most commonly used projection technique employed by engineers. CAD systems typically employ orthographic projection techniques, although the user is often provided with the option of changing to perspective projection.

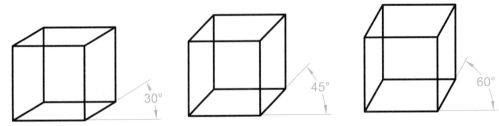

Figure 3-20 Receding axis angle

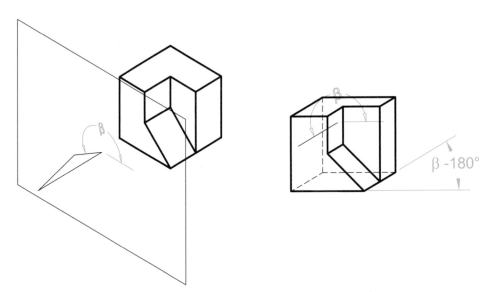

Figure 3-21 Relationship between the receding axis angle and in-plane projector angle β

Orthographic Projection Geometry

Orthographic projection is also a parallel projection technique, but differs from oblique projection in that the parallel projectors are perpendicular (normal) to the projection plane. Figure 3-22 shows a single projector and its orthogonal relationship to the projection plane.

Orthographic Projection Categories

Orthographic projections are subdivided according to the orientation of the object with respect to the projection plane. In an **axonometric projection**, all three principal axes are inclined to the projection plane. Figure 3-23 on page 61 shows this orientation, both as viewed from the front and from above. No axis is parallel (or perpendicular) to the projec-

tion plane. When projected, three principal faces of the object are visible. Axonometric projection results in an orthographic pictorial view.

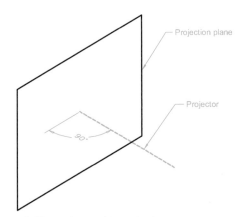

Figure 3-22 Orthographic projection geometry

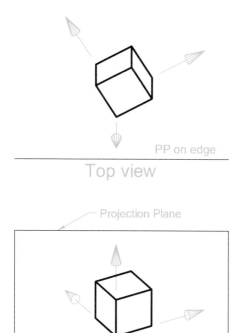

Figure 3-23   Object position in axonometric projection

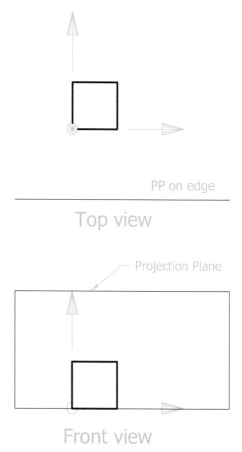

Figure 3-24   Object position in multiview projection

In a **multiview projection**, one object face and two principal axes are parallel to the projection plane. When projected, only one object face is visible. See Figure 3-24. In terms of the orientation of the object with respect to projection plane, multiview projection is identical to oblique projection, as well as one-point perspective projection.

□ AXONOMETRIC PROJECTIONS

Axonometric projections are classified according to the angles made by the principal axes when projected onto the projection plane. The upper half of Figure 3-25 on page 62 shows three different axonometric projections of a cube, along with an attached principal axis triad. In the lower portion of the figure only the projected axes are shown, and the angles between them.

In a **trimetric projection**, depicted on the left in Figure 3-25, none of the angles between the

projected principal axes are equal. The middle projection of Figure 3-25 is called a **dimetric projection**, where two of the three angles are equal. In an **isometric projection**, shown on the right in Figure 3-25, all three angles are equal.

Also note that, since all the equal-length axes depicted in Figure 3-25 are inclined to the projection plane, all of their projections are foreshortened. In the trimetric projection on the left all three axes are foreshortened by different amounts. Two of the three projected axes in the middle dimetric projection are foreshortened by the same amount. In the isometric projection on the right, there is an equal amount of foreshortening along all three axes.

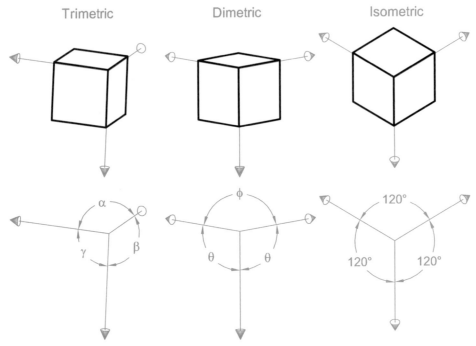

Figure 3-25 Axonometric projection classes

☐ ISOMETRIC PROJECTIONS

An isometric projection is foreshortened, or scaled, equally along all three principal axis directions. This fact makes isometric projections particularly useful in engineering. As a pictorial, an isometric projection is relatively easy to visualize, and it is also good at preserving the dimensional properties of the object.

To understand how an object must be oriented in order to obtain an isometric projection, imagine the projection of a cube. Figure 3-26 on the

left shows a trimetric view of the cube, on which a cube diagonal has been drawn. If the cube is rotated so that we look down the diagonal, as in Figure 3-26 on the right, we get an idea of how an isometric view is generated.

In most CAD systems an isometric view can be generated automatically. Figure 3-27 shows an example of a CAD viewing tool. Clicking on any

Figure 3-26 Looking down the diagonal of a cube

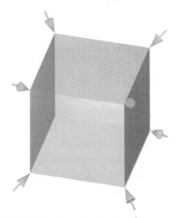

Figure 3-27 CAD system viewing tool

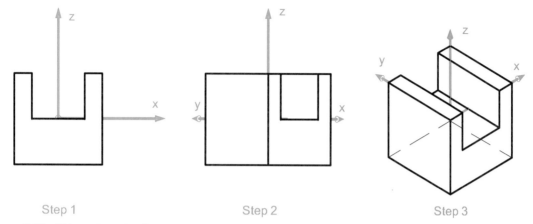

Figure 3-28 Object orientation for an isometric view

diagonal arrow generates one of eight possible isometric views.

More specifically, an isometric view is generated as shown in Figure 3-28:

1. Start with one principal face of the object parallel to the projection plane
2. Rotate the object about a vertical axis $(45 \pm 90n)$ degrees, where n is an integer
3. Rotate the object out of the horizontal plane by approximately $\pm 35.26°$[2]

Isometric Drawings

As a result of the particular object orientation in an isometric view, all three principal axes are foreshortened by exactly the same amount. This means that if a *solid model* is created in a CAD system, and an isometric view of the object is then printed out at a 1:1 scale, the printed (i.e., *projected*) edge lengths will all be equal to one another, but less than the actual edge lengths. It turns out that in an isometric projection, each principal axis is foreshortened to approximately 82% of its true length. For this reason, when plotting an isometric view in a CAD system, it is possible to correct for this foreshortening by multiplying the desired scale by the reciprocal of 0.82 (i.e.,~1.22).

[2] For a derivation of this value, see Ibrahim Zeid, *Mastering CAD/CAM*, McGraw-Hill, 2005, pp. 496–497.

When a sketch or drawing of an isometric view is made directly on paper, foreshortening effects are typically ignored. An isometric projection without foreshortening is referred to as an ***isometric drawing***. Note that, as shown in Figure 3-29, an isometric drawing is larger than a true isometric projection.

Multiview Projections

An arrangement for multiview projection is characterized by the following elements (see Figure 3-30). The:

1. projectors are parallel to one another
2. projectors are normal to the projection plane
3. object is positioned with one principal face parallel to the projection plane

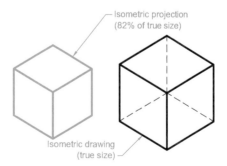

Isometric projection (82% of true size)

Isometric drawing (true size)

Figure 3-29 Comparison between an isometric projection and an isometric drawing

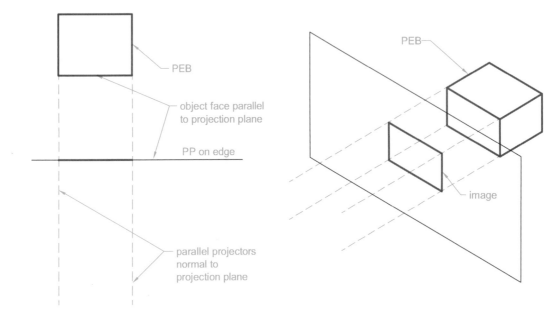

Figure 3-30 Multiview projection set up

As a direct result of these characteristics, multiview projections are good at preserving the object's dimensional information, but more than one view is necessary in order to fully describe the object.

In the following sections, different methods and techniques for constructing oblique and isometric pictorial sketches will be discussed. Multiview sketching is the subject of Chapter 4.

☐ INTRODUCTION TO PICTORIAL SKETCHING

In a pictorial view, three principal faces of an object are visible. A *pictorial sketch* shows an object's height, width, and depth in a single view. Unlike the multiview orthographic sketches discussed in the following chapter, a pictorial sketch conveys the object's three-dimensional shape in a single view. In this chapter, parallel projection (oblique and isometric) pictorials are discussed. Appendix A includes a discussion of perspective (both one-point and two-point) pictorials. Generally speaking, parallel projections preserve the object's metric properties, while perspective projections convey a strong sense of realism. Figure

3-6, reprinted on the page 65, shows how the same object would appear using these different pictorial projection techniques.

Regardless of the particular pictorial sketch being executed, the same general technique can be employed:

1. Using light construction lines, sketch a properly proportioned bounding box

2. Continuing with construction lines, add feature details

3. Starting with curved features, go bold

4. Complete sketch using bold lines.

This process is illustrated in Figure 3-31 on page 65 for an isometric pictorial.

Polyhedral shapes are frequently used for sketching. A *polyhedron* is a 3D solid bounded by a connected set of plane polygons, where every edge of a polygon belongs to just one other polygon. Polyhedral geometry consists of faces, edges, and vertices, as shown in Figure 3-32 on page 65. A *face* is a bounding surface on an object, whereas an *edge* serves as the intersection between two faces. Finally, a *vertex* is the endpoint of an edge.

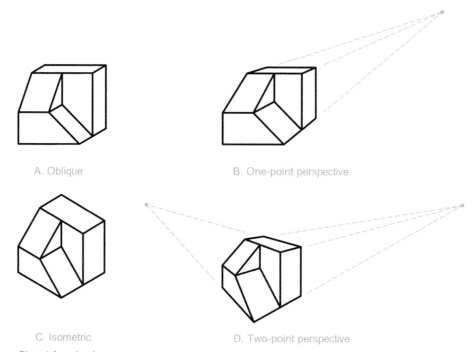

A. Oblique

B. One-point perspective

C. Isometric

D. Two-point perspective

Figure 3-6   Pictorial projections

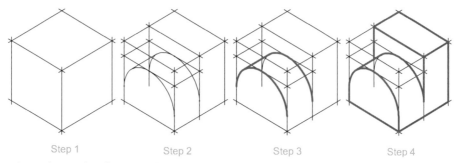

Step 1   Step 2   Step 3   Step 4

Figure 3-31   General procedure for pictorial (isometric) sketch

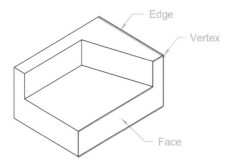

Edge

Vertex

Face

Figure 3-32   Cut block with vertex, edge, and face labeled

□ OBLIQUE SKETCHES

Introduction

An oblique sketch shows the true size and shape of one principal face of an object. In addition, two receding principal faces are also depicted in order to complete the pictorial view. Traditionally the object is positioned so that its most complex face (i.e., irregular, curved edges, etc.) is shown true size. Relating back to the discussion of oblique projections earlier in this chapter, the

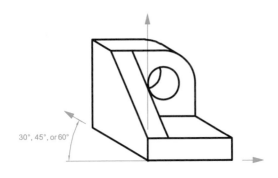

30°, 45°, or 60°

A. Receding axis up and to left

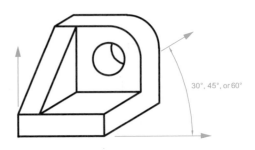

30°, 45°, or 60°

B. Receding axis up and to right

Figure 3-33   Oblique sketches of the same object

true size (typically front) face is oriented parallel to the projection plane.

Oblique sketches are relatively easy to draw. This is because complex features parallel to the projection plane project without distortion. For example, a circular edge will project as a circle, not an ellipse, as long as it is parallel to the projection plane.

Axis Orientation

In most oblique sketches, the object is oriented so that its front face is shown true size (parallel to the projection plane). Horizontal and vertical axes define the frontal plane. Width is measured along the horizontal axis, height along the vertical. The receding axis is then used to measure depth. The receding axis is drawn either up and to the left, or up and to the right. See Figure 3-33. The angle that the receding axis makes with the horizontal is usually drawn to be 30, 45, or 60 degrees.

Receding Axis Scale

Recall from the earlier section, Classes of Oblique Projections, that there are three different classes

Table 3-2   Classes of oblique sketches

Oblique Type	Receding Axis Scale
Cavalier	1
Cabinet	½
General	Between ½ and 1

of oblique pictorials: cavalier, cabinet, and general. The type of oblique is determined by the amount of scaling that occurs along the receding axis.[3] See Table 3-2.

In a cavalier oblique, the same scale factor is used along all three axes: horizontal, vertical, and receding. The resulting pictorial appears to be too long in the depth direction, due to the lack of foreshortening. See Figure 3-34 on page 67. To take into account this lack of foreshortening, a cabinet oblique employs a receding axis scale of one half. While convenient, as well as improving upon the appearance of a cavalier oblique, a cabinet oblique can appear to have too much foreshortening. To compensate for this in order to obtain a more pleasing pictorial, a general oblique that scales the receding axis between that of the cavalier and cabinet is sometimes used.

Object Orientation Guidelines

Two rules apply when selecting the front *true-size* face to be used in an oblique pictorial. The first rule has already been discussed, that is that the projection plane should be chosen so that it is parallel to the principal face containing the most complex (circular, curved) or irregular shape.

[3] Recall from the discussion of oblique projections earlier in the chapter that the amount of foreshortening along the receding axis is determined by the angle at which the parallel projectors pierce the projection plane. This angle is called the oblique projection angle (α). For a cavalier projection, α is 45 degrees; for a cabinet projection, α is ~64 degrees.

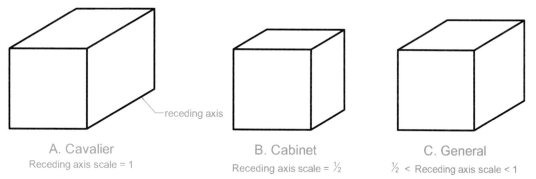

A. Cavalier	B. Cabinet	C. General
Receding axis scale = 1	Receding axis scale = ½	½ < Receding axis scale < 1

Figure 3-34   Cavalier, cabinet, and general oblique of a cube

This rule was employed, for example, in Figure 3-33, where the curved features are parallel to the projection plane. The second rule is to use the longest face as the front *true-size* view. Figure 3-35 shows two oblique pictorials of an angle beam, oriented according to the two rules.

In the event that the two rules are in conflict, the first rule takes precedence. In Figure 3-36,

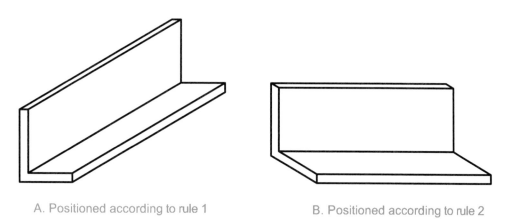

A. Positioned according to rule 1   B. Positioned according to rule 2

Figure 3-35   Object orientation for oblique pictorials; two rules

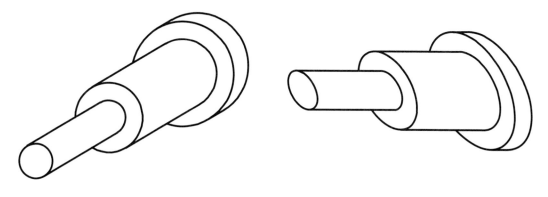

A. Rule 1 (preferred)   B. Rule 2 (avoid)

Figure 3-36   Object orientation for oblique pictorials; rule 1 preferred

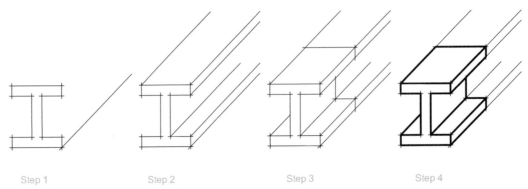

Step 1   Step 2   Step 3   Step 4

Figure 3-37   Oblique sketch of an extruded shape

pictorial A on the left is both easier to construct and less distorted than pictorial B.

Sketching procedure for a simple extruded shape

Drawing an oblique pictorial of an *extrusion* is particularly easy. See Figure 3-37.

1. Sketch the extruded profile as a front view; select and lightly sketch the direction and angle of receding axis
2. Sketch construction lines extending from other front face vertices, parallel to receding axis
3. Determine depth, then complete the back face of the object
4. Go bold.

Step-by-step cabinet oblique sketch example for a cut block (See Figure 3-38)

1. PEB construction.
   a. Identify the most *complex* (irregular shape, circular features, etc.) object face as the front view
   b. Lightly sketch a properly proportioned rectangle that just encloses object's front face
   c. Lightly sketch a receding axis (direction, angle) from one vertex of the front rectangle
   d. Mark off the appropriate depth along the receding axis (for cabinet, scale = ½)
   e. Complete the bounding box using construction lines
2. Using construction lines, block in feature details
3. Go bold.

Step-by-step cavalier oblique sketch example for object with circular features (See Figure 3-39 on page 69)

1. PEB construction.
   a. Identify the most *complex* (irregular shape, circular features, etc.) object face as the front view
   b. Lightly sketch a properly proportioned rectangle that just encloses object's front face
   c. Lightly sketch a receding axis (direction, angle) from one vertex of the front rectangle
   d. Mark off the appropriate depth along the receding axis (for cavalier, scale = 1)

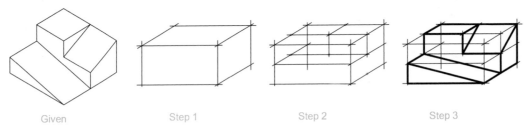

Given   Step 1   Step 2   Step 3

Figure 3-38   Multiple steps for cut block cabinet oblique sketch

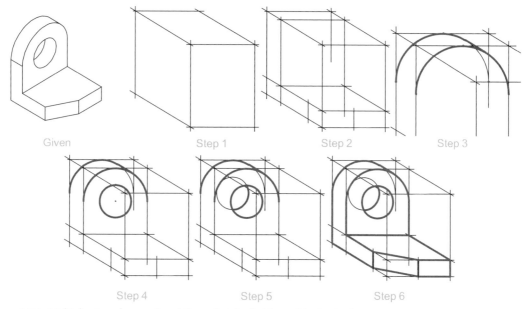

Figure 3-39 Multiple steps for cavalier oblique sketch of object with circular features

e. Still using construction lines, complete the bounding box

2. Using construction lines, block in linear features

3. Arc feature construction
   a. Locate arc quadrants (front and back)
   b. Sketch front face arc in bold
   c. Use construction lines to sketch back face arc
   d. Sketch a line tangent to the two arcs in bold
   e. Sketch visible portion of back arc in bold

4. Hole feature construction (front face)
   a. Locate center and quadrant points
   b. Sketch circle in bold

5. Hole feature construction (back face)
   a. Locate center and quadrant points
   b. Sketch circle using construction lines
   c. Sketch visible portion of back edge in bold

6. Go bold.

□ ISOMETRIC SKETCHES

Introduction

Similar to obliques, isometric pictorials are parallel projections from which dimensional information can be obtained. In addition, isometric views can easily be created in CAD systems, something not true of oblique projections.

Axis Orientation

In sketching an isometric, the principal axes are aligned as shown in Figure 3-40. One axis is vertical, while the other two are inclined at 30 degrees to the horizontal. Note that this 30-degree angle is a direct outcome of the object's orientation with respect to the projection plane used to generate an isometric projection, which has been described in the earlier section on isometric projections.

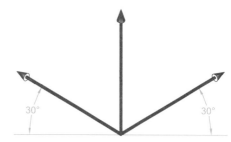

Figure 3-40 Isometric sketch axes

Isometric Scaling

Recall from the discussion earlier in the chapter that the most important property of an isometric projection is that all three principal axes are equally foreshortened. Therefore, for an isometric pictorial, all object edges parallel to a principal axis are scalable and can be directly measured.

For an isometric sketch, the foreshortening that would occur in a true isometric projection is generally ignored. Imagine, for example, an actual rectangular prism that measures $3 \times 2 \times 1$. An isometric sketch of this prism is shown in Figure 3-41. When creating the sketch the actual dimensions are laid out along the isometric axes, and not the foreshortened lengths.

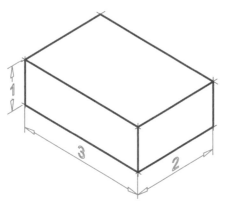

Figure 3-41   Isometric sketch of a rectangular prism

In an isometric sketch, object edges that are *not* parallel to a principal axis cannot be measured. In Figure 3-42, for example, the lengths of

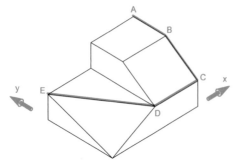

Figure 3-42   Isometric scaling

edges AB (parallel to the y axis) and CD (parallel to the x axis) can be directly measured. Lines parallel to a principal axis are referred to as **isometric lines**. Edge lengths BC and DE cannot be scaled (i.e., not measurable), since they are not parallel to any of the three principal axes. These are called **nonisometric lines**. In order to sketch these edges, their respective vertices must first be located.

Isometric Grid Paper

Isometric grid paper can be used when first learning to make isometric sketches. Figure 3-43 shows

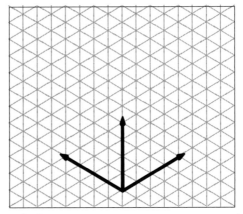

Figure 3-43   Isometric grid paper with superimposed isometric coordinate axes

an example of isometric grid paper, along with a superimposed isometric coordinate axes. Isometric grid paper consists of three sets of intersecting parallel lines; vertical, up and to the right at 30 degrees to horizontal, and up and to left at 30 degrees to horizontal. Figure 3-44 on page 71 shows an isometric sketch of a cut block using isometric grid paper.

Object Orientation Guidelines

Generally speaking, an object's longest principal dimension should appear as a horizontal dimension on the front face of the object. Assuming this to be the case, then an isometric view showing the

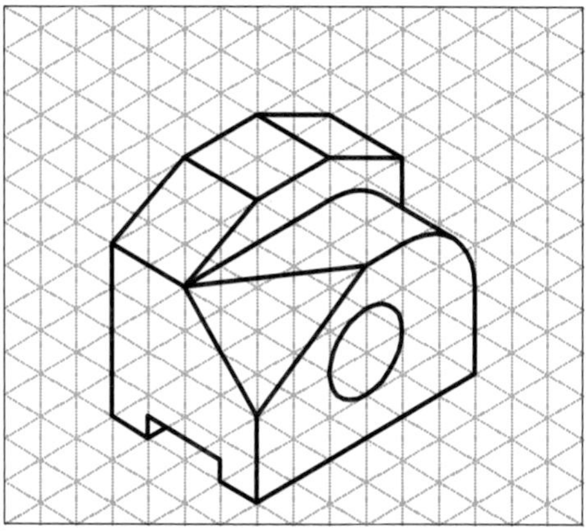

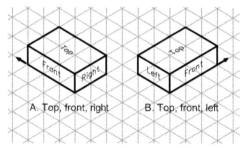

Figure 3-44 Isometric sketch of cut block using isometric grid paper

top, front and right side of an object is obtained by laying out the longest dimension on the leftmost *up and to the left* axis (see Figure 3-45A). Alternatively, to see a top, front, left isometric view, place the longest dimension along the rightmost *up and to the right* axis (Figure 3-45B).

Figure 3-45 Two isometric views of a rectangular prism

Step-by-step isometric sketch example for a cut block (See Figure 3-46)

1. PEB construction
   a. Identify the front view of the object to be sketched
   b. Using isometric axes (or isometric grid paper) lightly sketch a properly proportioned bounding box. To see the top, front, and right faces of the object, lay out the longest (horizontal) dimension along the *up and to the left* axis.
2. Still using construction lines, add feature details.
3. Go bold.

Circular Features in an Isometric View

In an isometric sketch a circular feature will appear as an ellipse. The following is a general procedure for sketching an isometric ellipse:

1. Identify the planar face on which the circular feature is to appear
2. Lightly sketch the principal axes of the feature on the planar face
3. Locate the quadrant points along these axes (equidistant from the intersection of the two axes)
4. If necessary, lightly sketch the bounding *rhombus* that just contains the circular edge
5. Using bold lines, sketch in the elliptical shape. The ellipse should pass through the quadrant points, and will be tangent to the rhombus sides.

Using this procedure, Figure 3-47 demonstrates the construction of an isometric cylinder.

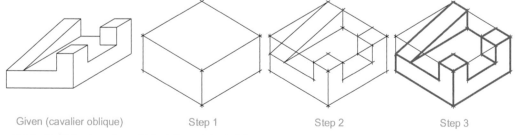

Given (cavalier oblique)   Step 1   Step 2   Step 3

Figure 3-46 Multiple steps for cut block isometric sketch

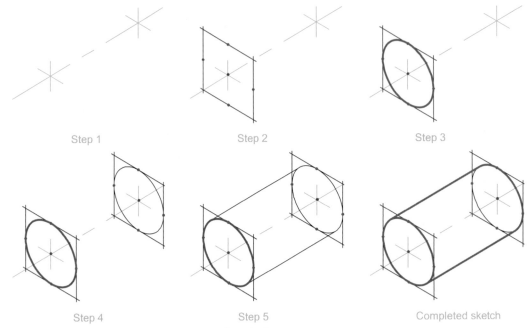

Step 1   Step 2   Step 3

Step 4   Step 5   Completed sketch

Figure 3-47   Construction of an isometric sketch of a cylinder

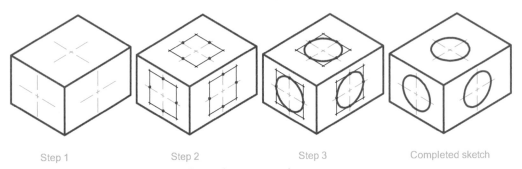

Step 1   Step 2   Step 3   Completed sketch

Figure 3-48   Construction of holes on three faces of an isometric box

In Figure 3-48, hole features are sketched on three faces of a box.

Step-by-step isometric sketch example for a cylinder (See Figure 3-47)

1. Sketch the axis, and locate the centers of the front and back faces of the cylinder. Note that, in this example, the cylinder axis is horizontal.

2. Locate the quadrant points on the front face of the cylinder, and then sketch in the bounding

rhombus. Note that the quadrant points are along the two axes in the plane of the face and equidistant from the center.

3. In bold, sketch the elliptical front face of the cylinder

4. Locate quadrant points, sketch the bounding rhombus, and lightly sketch the elliptical back face of the cylinder

5. Lightly sketch two lines to represent the *limiting elements* of the cylinder, parallel to the axis of the cylinder and tangent to the front and back elliptical faces

6. Go bold, showing only the visible portion of the back edge.

Step-by-step isometric sketch example for a box with holes on three faces (See Figure 3-48 on page 72)

1. Locate the centers of the holes on each of the three visible faces of the box
2. Locate the quadrant points of the holes, then sketch in the bounding rhombus
3. Go bold.

Step-by-step sketch example for object with circular features (See Figure 3-49)

1. PEB construction
   a. Construct two boxes
2. Locate hole centers
   a. Upper-left through hole
   b. Lower-right arc and through hole
3. Locate top face quadrants, rhombus construction
   a. Locate quadrants for upper-left through hole
   b. Use construction lines to create rhombus
   c. Locate quadrants for lower-right arc
4. Top face ellipse construction
   a. Sketch upper-left full ellipse in bold
   b. Sketch lower-right partial ellipse in bold
   c. Locate quadrants for lower-right through hole
   d. Use construction lines to create rhombus
5. Top face ellipse construction continued
   a. Sketch lower-right full ellipse in bold
6. Bottom face ellipse construction
   a. Upper-left full ellipse using light construction lines
   b. Lower-right full ellipse using light construction lines
   c. Lower-right partial ellipse using light construction lines
7. Go bold.

Chapter review: pictorial sketching scalability

A drawing is said to be *scalable* if dimensional information can be derived from it, even if the drawing itself is not to scale. If, for example, the actual measure of a certain distance shown on the drawing is known, then other dimensions on the

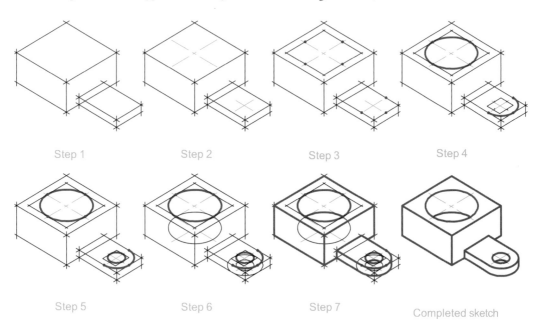

Step 1   Step 2   Step 3   Step 4

Step 5   Step 6   Step 7   Completed sketch

Figure 3-49 Multiple steps for isometric sketch of object with circular features

drawing can be approximated by forming proportional relationships between the actual and the measured distances:

$$\frac{x_{actual}}{y_{actual}} = \frac{x_{measured}}{y_{measured}}$$

Multiview

- Parallel projection technique
- One PEB face is parallel to the projection plane

Therefore all edges that are parallel to the projection plane are scalable.

Oblique

- Parallel projection technique
- One PEB face is parallel to the projection plane

Therefore all edges that are parallel to the projection plane are scalable.

- For cavalier oblique, the receding axis is scaled the same as the other principal axes

Therefore all edges parallel to the receding axis are scalable on a cavalier oblique.

Isometric

- Parallel projection technique
- PEB oriented such that all three principal axes are equally foreshortened

Therefore all edges parallel to any principal axis are scalable.

Trimetric

- Parallel projection technique
- PEB oriented such that all three principal axes are foreshortened by different amounts

Therefore all edges parallel to a single principal axis are scalable. In effect, there are three separate scales in a trimetric projection, one for each principal axis.

NOTE: For any planar projection technique, if an object (or feature) is parallel to the projection plane, the feature will be projected true shape. For a parallel projection, these features will also be projected true size.

☐ QUESTIONS

TRUE AND FALSE

1. Parallel object edges always appear parallel in a parallel projection.
2. In a trimetric projection, none of the angles between the projected principal axes are equal.
3. Assuming that a projection plane is offset to a new parallel position, an axonometric projection of an object onto the projection plane will be identical (i.e., same size and shape) in either location.
4. In an isometric drawing, each principal axis is foreshortened by approximately 82% of its true length.
5. The receding axis angle of an oblique projection is governed by the out-of-plane angle α.

MULTIPLE CHOICE

6. Which of the following is not considered to be a main element of a projection system?
   a. 3D object
   b. 2D cutting plane
   c. Projectors
   d. 2D projection plane
   e. 2D projected image

7. For the cavalier oblique drawing of the cut block shown in Figure P3-1, which edges are scalable (i.e., directly measurable)?
   a. BC & AC
   b. EF & DJ
   c. FH & HJ
   d. GJ & DJ
   e. All of the previous
   f. None of the previous

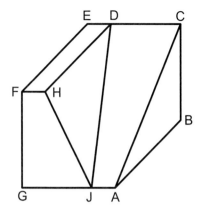

Figure P3-1   Cavalier oblique drawing of a cut block

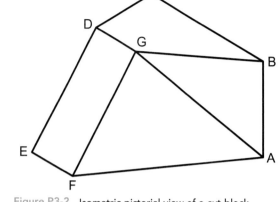

Figure P3-2   Isometric pictorial view of a cut block

8. For the isometric drawing of the cut block shown in Figure P3-2, which edges are scalable (i.e., directly measurable)?

   a. BG

   b. AG

   c. AF

   d. DE

   e. All of the previous

   f. None of the previous

9. Referring to Figure P3-3, if the sight lines are parallel to each other and also perpendicular to the infinite projection plane, $\beta = 15°$, and $\phi = 45°$, what is the type of projection of the block that will appear on the projection plane?

   a. One-point perspective

   b. Two-point perspective

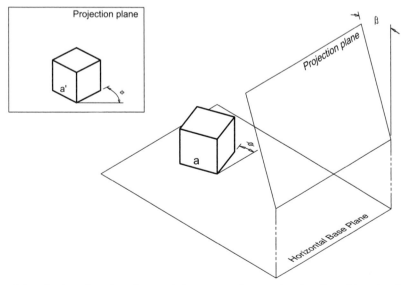

Figure P3-3   Pictorial projection where lengths a and a′ represent the actual and projected lengths of the block, respectively (Figure adapted from the work of Michael H. Pleck).

c. Trimetric

d. Isometric

e. Dimetric

f. Cabinet oblique

g. Cavalier oblique

h. General oblique

i. None of the previous

10. Referring to Figure P3-3, if the sight lines are perpendicular to the infinite projection plane, $\beta = 30°$ and $\phi = 20°$ initially, which of the following is true of the ratio between the projected length a' and the actual length a as ϕ is increased to 70°?

a. Increases

b. Decreases

c. Does not change

d. Cannot determine without additional information

SKETCHING

11. Given the isometric view of the cut block objects appearing in P3-4 through P3-65, use Dwg3-2 (or download worksheet from the book Web site) to sketch a cavalier oblique view of the object.

12. Given the isometric view of the cut block objects appearing in P3-4 through P3-65, use

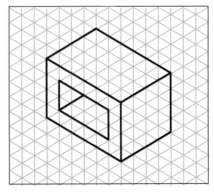

Figure P3-5

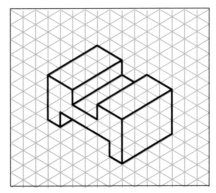

Figure P3-6

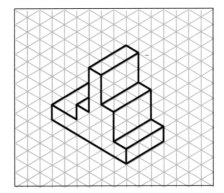

Figure P3-7

Dwg3-2 (or download worksheet from the book Web site) to sketch a cabinet oblique view of the object.

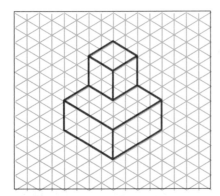

Figure P3-4

13. Given the isometric view of the cut block objects appearing in P3-4 through P3-65, use Dwg3-3 (or download worksheet from the book Web site) to sketch a cavalier oblique view of the object.

14. Given the isometric view of the cut block objects appearing in P3-4 through P3-65, use Dwg3-3 (or download worksheet from the book Web site) to sketch a cabinet oblique view of the object.

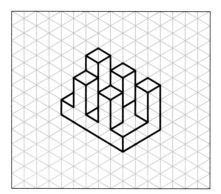

Figure P3-8

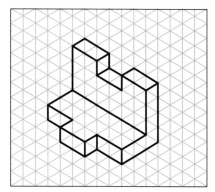

Figure P3-9

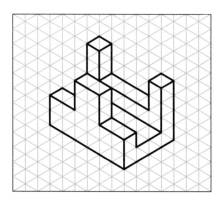

Figure P3-10

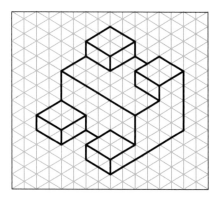

Figure P3-11

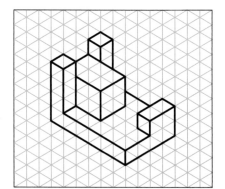

Figure P3-12

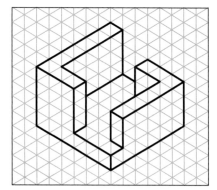

Figure P3-13

15. Given the isometric view of the cut block objects appearing in P3-4 through P3-65, use a blank sheet of paper to create a well-proportioned cavalier oblique sketch of the object.

16. Given the isometric view of the cut block objects appearing in P3-4 through P3-65, use a blank sheet of paper to create a well-proportioned cabinet oblique sketch of the object.

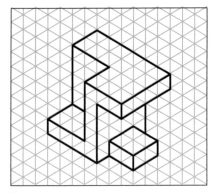

Figure P3-14

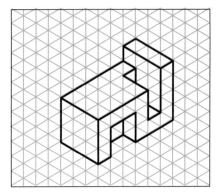

Figure P3-15

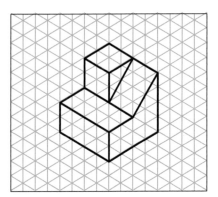

Figure P3-16

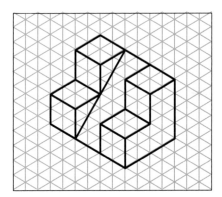

Figure P3-17

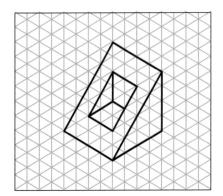

Figure P3-18

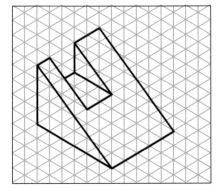

Figure P3-19

17. Given the cavalier oblique view of the cut block objects appearing in P3-66 through P3-95, use Dwg3-1 (or download worksheet from the book Web site) to sketch an isometric view of the object.

18. Given the cabinet oblique view of the cut block objects appearing in P3-66 through P3-95, use Dwg3-1 (or download worksheet from the book Web site) to sketch an isometric view of the object.

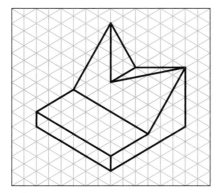

Figure P3-20

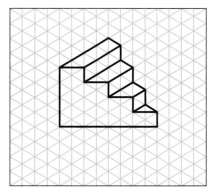

Figure P3-21

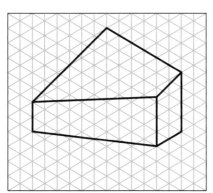

Figure P3-22

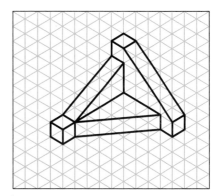

Figure P3-23

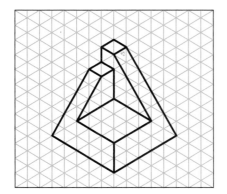

Figure P3-24

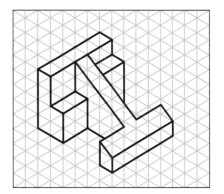

Figure P3-25

19. Given the cavalier oblique view of the cut block objects appearing in P3-66 through P3-95, use a blank sheet of paper to create a well-proportioned isometric sketch of the object.

20. Given the cabinet oblique view of the cut block objects appearing in P3-66 through P3-95, use a blank sheet of paper to create a well-proportioned isometric sketch of the object.

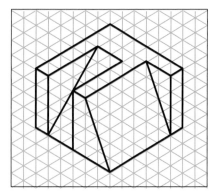

Figure P3-26

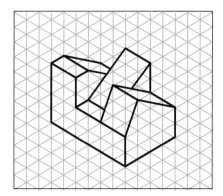

Figure P3-27

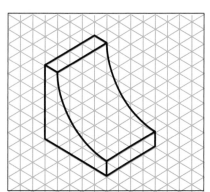

Figure P3-28

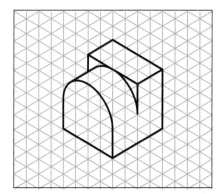

Figure P3-29

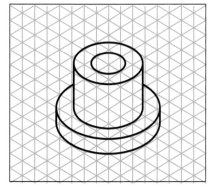

Figure P3-30

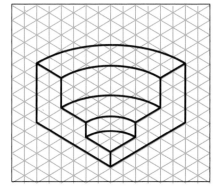

Figure P3-31

21. Use Dwg3-4 (or download worksheet from the book Web site) to rotate and sketch the given objects (A, B, or C) so that the labeled edge xy corresponds to the new orientation:
    A. Given cabinet oblique, sketch isometric
    B. Given cavalier oblique, sketch isometric
    C. Given isometric, sketch cavalier oblique

22. Use Dwg3-5 (or download worksheet from the book Web site) to rotate and sketch

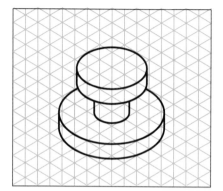

Figure P3-32

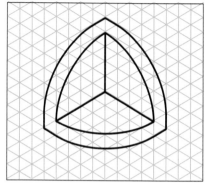

Figure P3-33

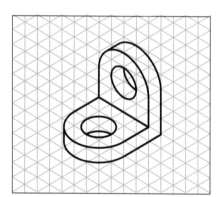

Figure P3-34

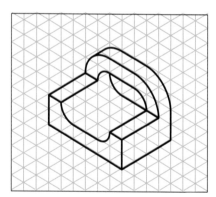

Figure P3-35

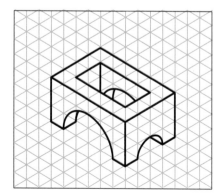

Figure P3-36

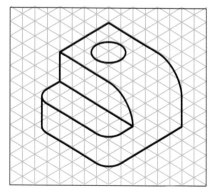

Figure P3-37

the given objects (A, B, or C) so that the labeled edge xy corresponds to the new orientation:

a. Given isometric, sketch cabinet oblique
b. Given cavalier oblique, sketch isometric
c. Given cavalier oblique, sketch isometric

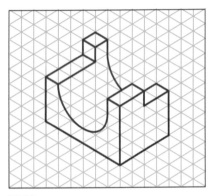

Figure P3-38

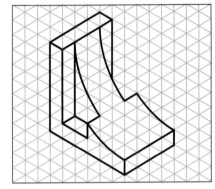

Figure P3-39

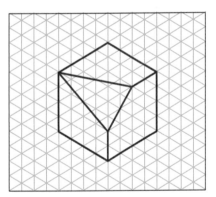

Figure P3-40

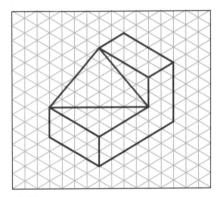

Figure P3-41

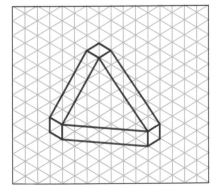

Figure P3-42

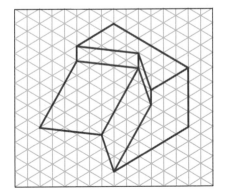

Figure P3-43

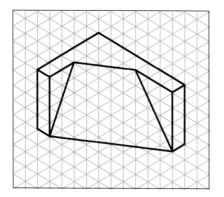

Figure P3-44

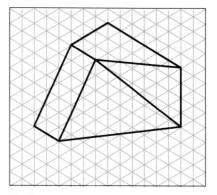

Figure P3-45

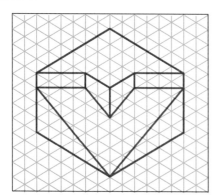

Figure P3-46

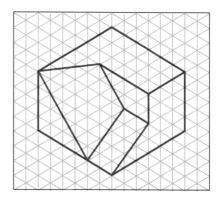

Figure P3-47

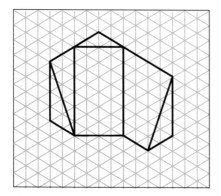

Figure P3-48

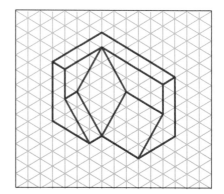

Figure P3-49

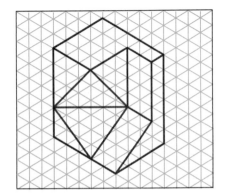

Figure P3-50

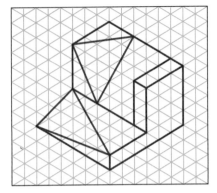

Figure P3-51

Figure P3-52

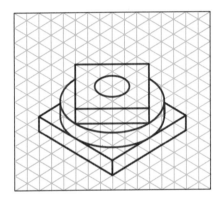

Figure P3-53

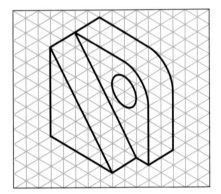

Figure P3-54

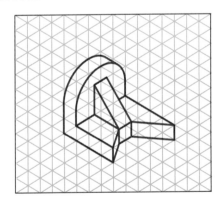

Figure P3-55

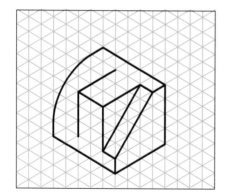

Figure P3-56

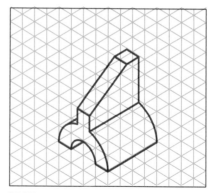

Figure P3-57

Figure P3-58

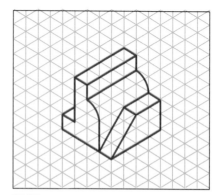

Figure P3-59

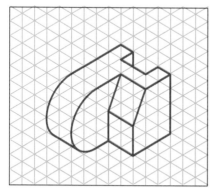

Figure P3-60

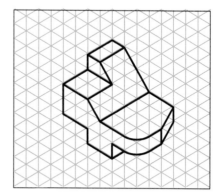

Figure P3-61

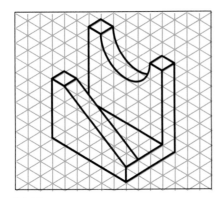

Figure P3-62

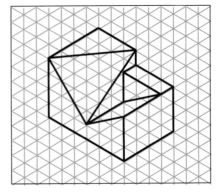

Figure P3-63

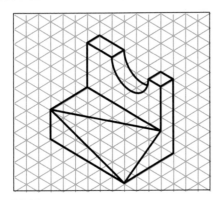

Figure P3-64

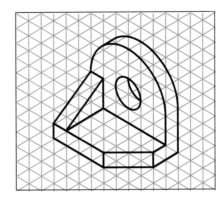

Figure P3-65

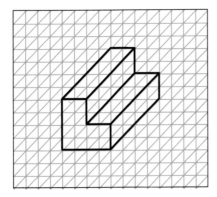

Figure P3-66

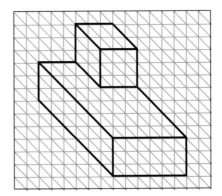

Figure P3-67

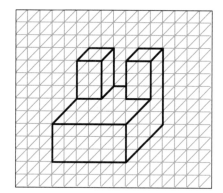

Figure P3-68

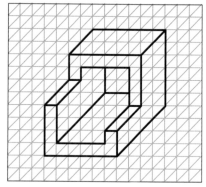

Figure P3-69

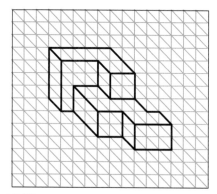

Figure P3-70

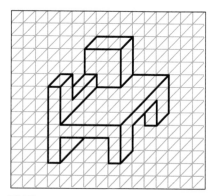

Figure P3-71

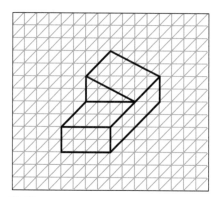

Figure P3-72

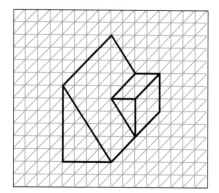

Figure P3-73

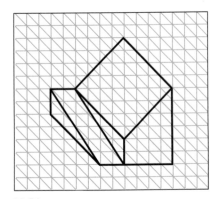

Figure P3-74

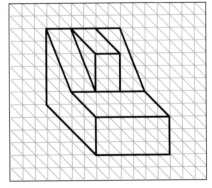

Figure P3-75

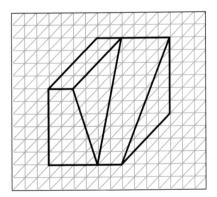

Figure P3-76

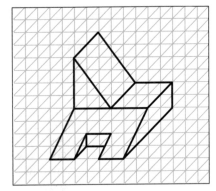

Figure P3-77

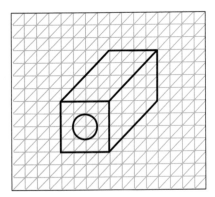

Figure P3-78

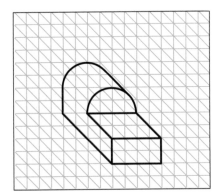

Figure P3-79

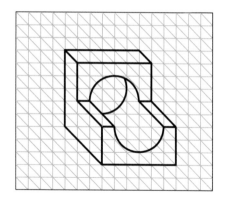

Figure P3-80

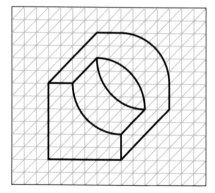

Figure P3-81

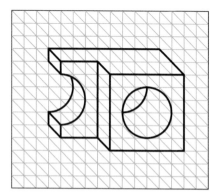

Figure P3-82

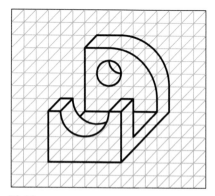

Figure P3-83

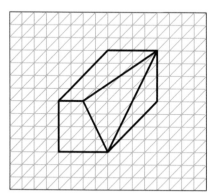

Figure P3-84

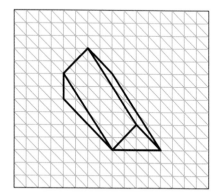

Figure P3-85

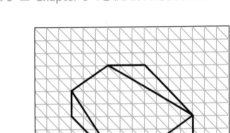

Figure P3-86

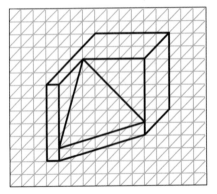

Figure P3-87

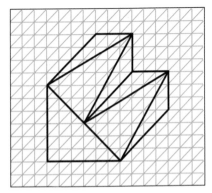

Figure P3-88

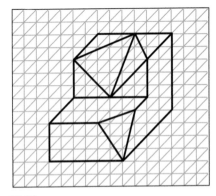

Figure P3-89

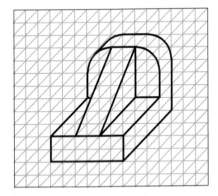

Figure P3-90

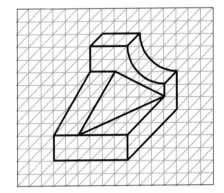

Figure P3-91

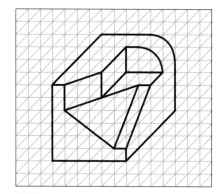

Figure P3-92

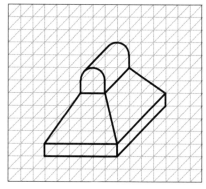

Figure P3-93

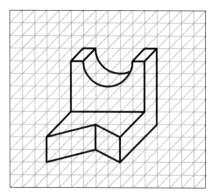

Figure P3-94

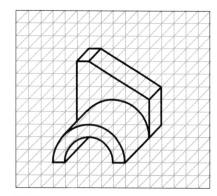

Figure P3-95

4 MULTIVIEWS

☐ MULTIVIEW SKETCHING

Introduction—Justification and Some Characteristics

Multiview drawings are at the core of what has traditionally been thought of as engineering graphics. The purpose of a multiview drawing is to represent fully the size and shape of an object using one or more views. Along with notes and dimensions, these views provide the information needed to fabricate the part.

Chapter 3 included a brief discussion of the characteristics of multiview projection. These characteristics, as seen in Figure 4-1, include 1) parallel projectors normal to the projection plane and 2) the object positioned so that one principal face is parallel to the projection plane.

As a consequence of this geometry, a multiview drawing can show only one object face. This means that in most cases more than one view is needed to describe the object fully. It is for this reason that this orthographic projection technique is called multiview projection.

While only two of the three sets of linear dimensions (i.e., width, depth, height) are projected in any one view, all of this projected information parallel to the projection plane is directly scalable.

Glass Box Theory

Since more than one view is typically needed to document an object using multiview projection, **glass box theory** is used to describe the arrangement of the different multiviews with respect to one another. Imagine that the object to be documented is placed inside a glass box, as shown in Figure 4-2. The object is positioned so that its sides are orthogonal to the sides of the glass box. Notice also that the principal dimensions of the object, width, depth and height, are also indicated in this figure.

In Figure 4-3 on page 93, the six sides of the glass box are used as projection planes, upon

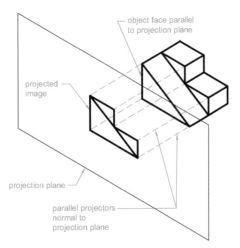

Figure 4-1 Pictorial view of the multiview projection process

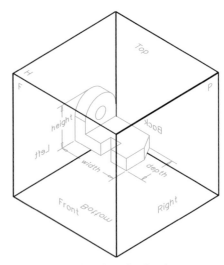

Figure 4-2 Object placed inside glass box

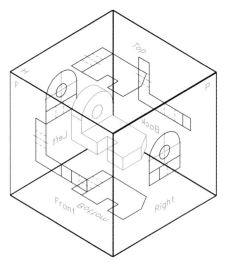

Figure 4-3   Object projected onto box sides

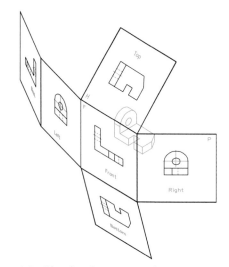

Figure 4-4   Glass box being opened

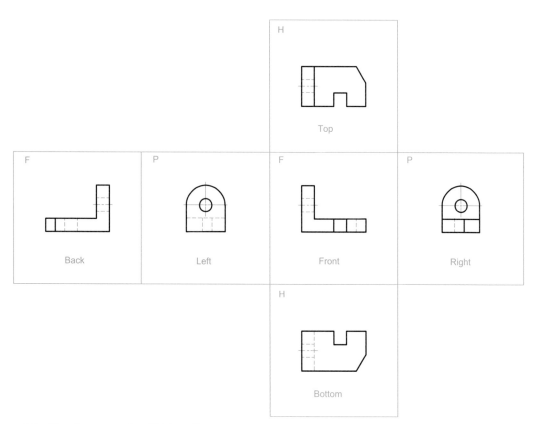

Figure 4-5   Glass box opened and laid out flat

which the six principal views (top and bottom, front and back, right and left) of the object are projected.

Imagine further that some of the glass box sides are hinged to one another. These hinge, or **fold lines**, when opened (as shown in Figure 4-4 on page 93) and then laid flat, result in the six views being arranged as shown in Figure 4-5 on page 93.

Notice in Figure 4-5 that four of the views are hinged to the front view, which is traditionally treated as the primary view in multiview projection. Also notice that the top, front and bottom views are all vertically aligned while the back, left, front, and right views are horizontally aligned.

All six principal views are not normally required to document an object completely. Notice in Figure 4-5 the similarities between top and bottom, front and back, left and right. Three principal views are sufficient to fully describe most objects. Most commonly these views are top, front, and right, as seen in Figure 4-6.

In line with this, we normally speak of three (not six) mutually perpendicular projection planes: Horizontal (H), Frontal (F), and Profile (P). The top and bottom views are projected onto H, front and back onto F, left and right onto P.

Alignment of Views

Multiviews are always aligned according to the dictates of the glass box projection planes and

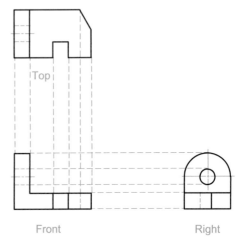

Figure 4-7   Feature alignment

their fold lines. As can be seen in Figure 4-7, not only the extents but also the internal features should be aligned.

Figure 4-8 shows that the top and front views are vertically aligned and they share the width dimension, while the front and right views are horizontally aligned and share the height dimension. Aligned views that share a common dimension are said to be **adjacent**. The top and right views, while not aligned, share the depth dimension. Views that share a common dimension, but are not aligned, are said to be **related**.

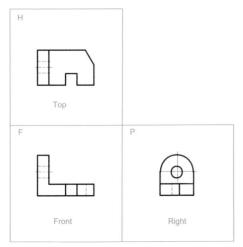

Figure 4-6   Top, front, right multiview arrangement

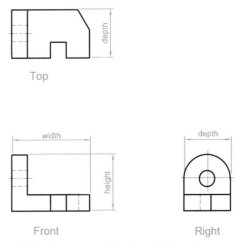

Figure 4-8   Three views with shared dimensions

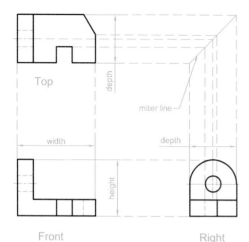

Figure 4-9   Transfer of depth using a miter line

Transfer of Depth

Every point or feature appearing in one view must be aligned along parallel projectors in their adjacent and related views. Between adjacent views feature information can be transferred directly along parallel projectors (see Figure 4-7 on page 94). Either a trammel or a 45-degree miter line can be used to transfer depth information between related views (see Figure 4-9).

View Selection

The most descriptive view should be selected as the front view. In addition, the longest principal dimension should appear as a horizontal dimension in the front view. For example, in the multiview drawing of the boat in Figure 4-10, the side of the boat appears in the front view, because it is the most descriptive view and the longest dimension appears in the front view as horizontal.

Another guideline related to view selection is that the minimum number of views should be included that allows a complete, unambiguous representation of the object. For most objects, three views are required to document the part fully. In some cases, however, only two views are needed (see Figure 4-11 on page 96), simple parts (e.g., washers, bushings) may only require a single view, along with a dimensional callout (see Figure 4-12 on page 96).

In the event that two views provide the same information, choose the view that has the fewest number of hidden lines. In Figure 4-13 on page 97 the right view is preferable to the left view because the right view has fewer hidden lines.

Third- and First-Angle Projection

Using a horizontal and a (vertically oriented) frontal plane, three-dimensional space can be

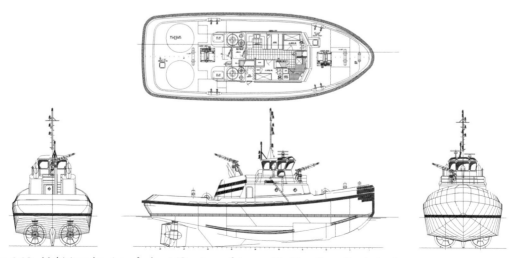

Figure 4-10   Multiview drawing of a boat (Courtesy of Jensen Maritime Consultants, Inc.)

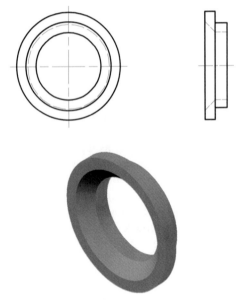

Figure 4-11 Two view drawing

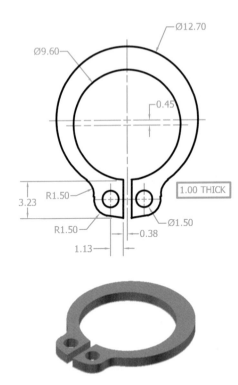

Figure 4-12 One view drawing

divided into quadrants, numbered as shown in Figure 4-14 on page 97.

Third-angle projection, used in the United States and Great Britain, assumes that an object to be projected resides in the third quadrant. Using the viewing directions for top, front and right shown in Figure 4-14, the object can be isolated as shown in Figure 4-15A on page 97. Note that in third-angle projection, the projection planes are between the viewer and the object. After projecting the views and opening the projection planes, Figure 4-15B results. Figure 4-15C shows the resulting view arrangement employed in third-angle projection.

In first-angle projection, used in the rest of Europe and Asia, the object to be projected is placed in the first quadrant (see Figure 4-14). Using the same viewing directions, the object and its projection planes have been isolated as shown in Figure 4-16A on page 97. Note that in first-angle projection the projection planes are all placed behind or below the object. After projecting and then unfolding the projection planes, Figure 4-16B results.

To indicate whether third- or first-angle projection has been used, the symbols shown in Figure 4-17 on page 98 are used.

Line Conventions

In a drawing view, dark thick *continuous lines* are used to represent the

1. Edge view of a surface
2. Edge between two intersecting surfaces
3. Extent of a contoured surface

Examples of each of these are shown in Figure 4-18 on page 98.

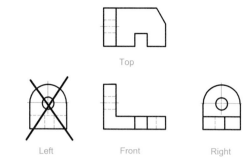

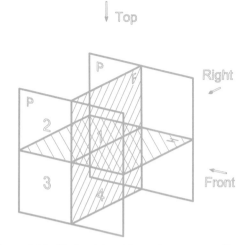

Figure 4-13   Select the view with fewer hidden lines

Figure 4-14   Horizontal and front plane, two profile planes

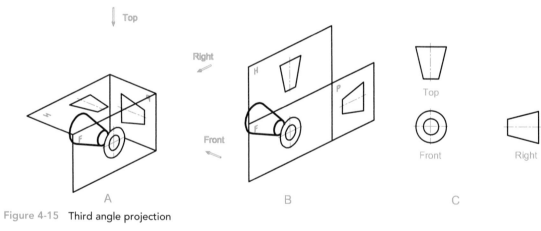

Figure 4-15   Third angle projection

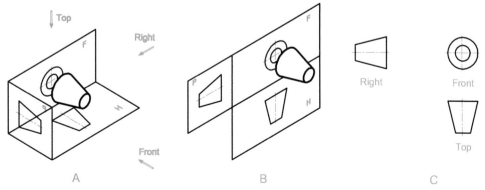

Figure 4-16   First angle projection

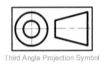

Third Angle Projection Symbol   First Angle Projection Symbol

Figure 4-17 Third and first angle projection symbols

Dark thin dashed lines, called ***hidden lines***, are used to represent features that are hidden in a particular view. Similar to visible continuous lines, hidden lines are used to represent:

1. A hidden edge of a surface
2. A hidden change of planes
3. The hidden extents or limiting elements of a hole

See Figure 4-19 for examples of each of these occurrences.

Centerlines are used in a variety of situations. These thin dark lines typically extend about 5 millimeters beyond the feature being repre-sented. Centerlines are commonly used to repre-

sent the axis of a cylinder or hole. In a circular view crossing centerlines are used. These center-lines should extend beyond the largest diameter (or radius) concentric circle (or arc) being repre-sented. In the rectangular view a single centerline represents the axis of the cylinder or hole. See Figure 4-20 for an example.

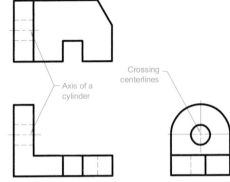

Figure 4-20 Centerline usages

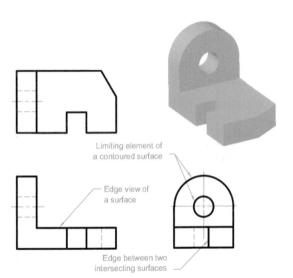

Figure 4-18 Visible continuous line usages

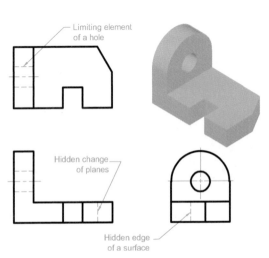

Figure 4-19 Hidden line usages

Multiview drawing of a cylinder

A solid cylinder has two circular edges. In the rectangular view these edges project as straight lines. In order to complete the representation, the *limiting elements*, or extents, of the cylinder are also represented as continuous lines. See Figure 4-21.

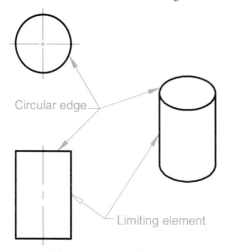

Circular edge

Limiting element

Figure 4-21 Multiview drawing of a cylinder

Centerlines are also used to show a path of motion, indicate symmetry, or represent bolt circles. See Figure 4-22.

Line Precedence

Different object features may sometimes coincide in a multiview sketch. When this occurs, the following order of line precedence is used to determine which lines are represented and which are not: (1) visible, (2) hidden, (3) center. Figure 4-23 on the facing page illustrates three different collinear line combinations.

Generic three multiview sketch procedure

1. Using light construction lines, sketch three properly proportioned bounding box views
2. Add front view feature details

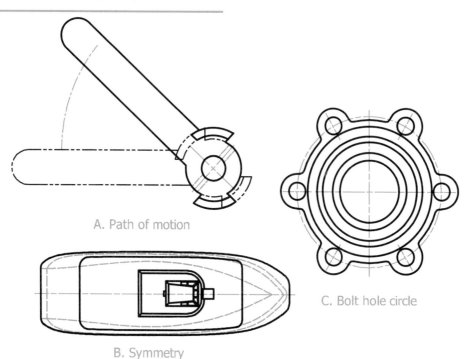

A. Path of motion

B. Symmetry

C. Bolt hole circle

Figure 4-22 Other centerline usages

visible over center

hidden over center

visible over hidden

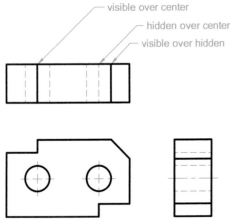

Figure 4-23 Line precedence

3. Project feature details from front view to adjacent views

4. Starting with curved features, go bold

See Figure 4-24.

Step-by-step multiview sketch example

1. Using light construction lines, sketch three properly proportioned bounding box views

2. Add visible edge details in top and front views

3. Project feature details from front and top views to left view; use miter line or trammel to transfer depth information from top to left view. Also project additional feature details between top and front views.

4. Using light construction lines, layout left side view, as well as hidden lines in other views

5. Starting with curved features, go bold

See Figure 4-25 on page 101.

Intersections and Tangency

When a planar surface intersects a curved surface, a line is drawn to represent the intersecting

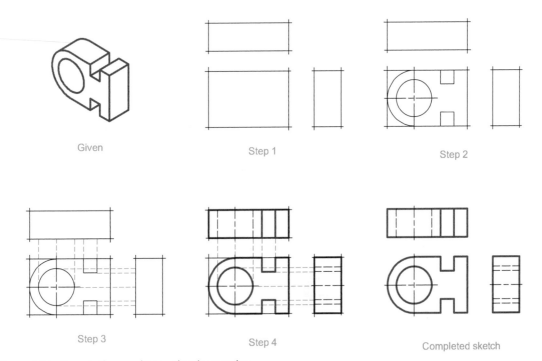

Given

Step 1

Step 2

Step 3

Step 4

Completed sketch

Figure 4-24 Generic three multiview sketch procedure

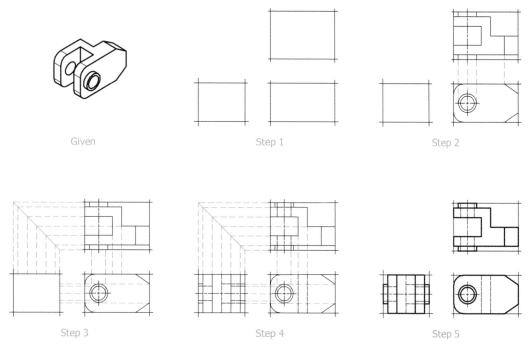

Given   Step 1   Step 2

Step 3   Step 4   Step 5

Figure 4-25   Step-by-step multiview sketch example

surfaces. In the event that the planar surface is tangent to a curved surface, no edge is shown to represent the line of tangency. See Figure 4-26 on page 102.

Fillets and Rounds

In designing parts, sharp corners are avoided. Not only are they difficult to fabricate, but they can also lead to stress concentrations, resulting in weakened parts. A ***fillet*** is used to eliminate an internal corner, while a ***round*** removes an external corner. See Figure 4-27 on page 102.

Cast parts are designed with fillets and rounds. Due to the nature of this manufacturing process, cast parts have rough external surfaces. In order to mate a casting with another part, it is often necessary to machine the original surfaces in order to create a good mating surface. For this reason a cast part with rounded corners generally indicates that the part is unfinished, while sharp corners indicate that the surface has been machined. See Figure 4-28 on page 103.

Fillet and round features are displayed as small arcs in a multiview drawing, as shown in Figure 4-29 on page 103. In their rectangular views, these fillet features are typically not shown.

A fillet or round connects two otherwise intersecting surfaces with a curved surface tangent to the original surfaces. Since there is no real change in planes, the top view of the object shown in Figure 4-30 on page 104 would normally be shown as a single surface (see Figure 4-30A). In order to provide a clearer representation, however, fillets and rounds are sometimes ignored in the rectangular view; edges are then drawn at the imaginary intersection of the two planes, as seen in Figure 4-30B.

Machined Holes

Machined holes are formed by various machining operations. These include drilling, boring, and reaming. The specific machining operation used to create the hole is not specified on the drawing, leaving this decision to the machinist. The diameter of a hole, not the radius, is specified,

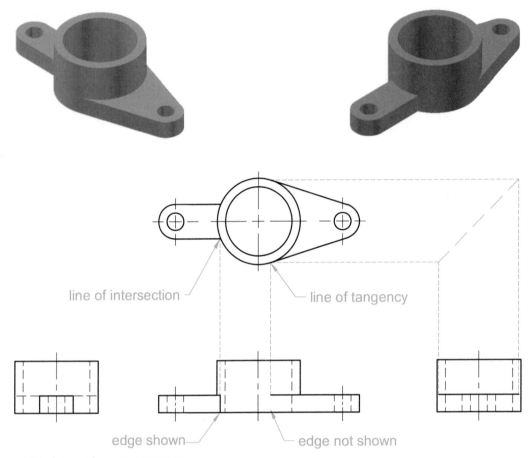

line of intersection

line of tangency

edge shown

edge not shown

Figure 4-26 Intersection versus tangency

using a leader extending from the circular view. Figure 4-31 on page 104 shows several different kinds of machined holes. A ***through*** hole, formed

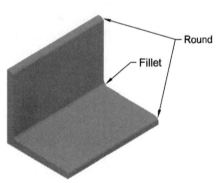

Round

Fillet

Figure 4-27 Fillets and rounds

by drilling, goes all the way through the part. A ***blind*** hole, on the other hand, has a specific depth. Since a blind hole is also formed by drilling, the bottom of the hole comes to a conical point formed by the drill bit. Only the cylindrical portion of the hole should be dimensioned. The angle of the drill bit is 30 degrees. A ***counterdrilled*** hole is formed by drilling a larger hole inside a smaller hole in order to enlarge the initial portion of the hole. As seen in Figure 4-31, a 120-degree shoulder is a byproduct of the counterdrill operation. The process of drilling and then conically enlarging a hole is called ***countersinking***. A countersunk hole is used for flat head fasteners, and may also serve as a chamfered guide for shafts and other cylindrical parts. In a countersunk hole both the diameter and the angle of the countersink are

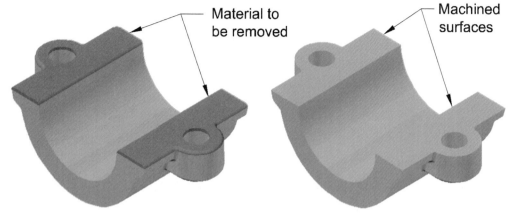

Figure 4-28 Cast part surfaces before and after machining

specified. Although the angle of the countersink is typically 82 degrees, by convention it is often drawn as 90 degrees. **Spotfacing** refers to the process of machining the surface around a drilled hole, typically on a cast part, in order to provide a smooth mating surface for washers, bolt heads, nuts, etc. The cylindrical diameter created by the spotfacing operation is specified, and the required depth is left to the machinist. **Counterboring** refers to the process of cylindrically enlarging the initial portion of a drilled hole. The counter-

bore operation results in an enlarged hole with a flat bottom. A counterbore hole permits a bolt head to be flush with or recessed below the surface of the part. In a **threaded** hole, an internal thread is made by drilling a hole with a tap drill.

Conventional Representations: Rotated Features

A true orthographic projection of a part with radially distributed features like ribs, holes, spokes, etc. can be confusing to visualize and difficult to construct. The front view of Figure 4-32A on page 105, for example, shows the true projection of a part with radially distributed ribs and holes. Notice the lack of symmetry about the centerline. The rib(s) on the left side of Figure 4-32A will not be easy to draw, since it is not parallel to the frontal plane. Also, the holes are not symmetrical about the centerline.

To avoid this situation, by convention these views are simplified by rotating the radial features so that they are aligned in a single plane that is perpendicular to the line of sight. Looking at the top view of Figure 4-32B, imagine that a rib and a hole are rotated onto the horizontal axis, as indicated by the arrows. The front view of Figure 4-32B shows this conventional representation, with the revolved features now aligned.

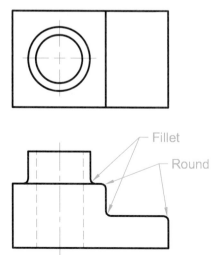

Figure 4-29 Fillet features display as arcs

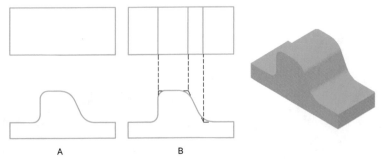

Figure 4-30 **Fillet conventions**

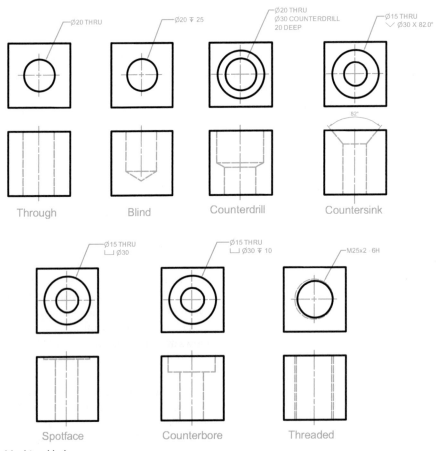

Figure 4-31 **Machined holes**

A. True projection   B. Conventional projection

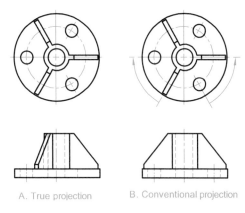

Figure 4-32   Treatment of revolved features

Step-by-step multiview sketch example: object with complex features

1. Using light construction lines, sketch three properly proportioned bounding box views
2. Add front view feature details
3. Project feature details from front view to adjacent views
4. Starting with curved features, go bold

See Figure 4-33.

□ VISUALIZATION TECHNIQUES FOR MULTIVIEW DRAWINGS

Introduction and Motivation

Visualization is a process by which shape information on a drawing is translated to give the viewer an understanding of the object represented. The part shown in the multiview drawing in Figure 4-34 on page 106, for example, may not be easily recognizable, even to an experienced engineer. It is only after a careful reading of the

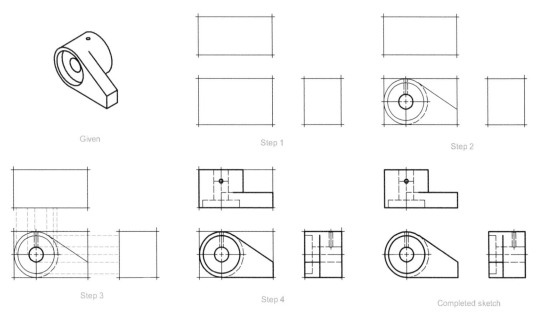

Given   Step 1   Step 2

Step 3   Step 4   Completed sketch

Figure 4-33   Step-by-step multiview sketch example: object with complex features

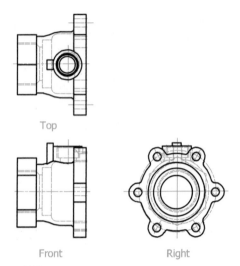

Top

Front   Right

Figure 4-34 Multiview drawing of a complex part

drawing that a mental image of the product begins to emerge (see Figure 4-35). In the remaining sections of this chapter various spatial visualization techniques will be discussed.

Treatment of Common Surfaces

NORMAL SURFACES

A rectangular prism like the one depicted in Figure 4-36 only contains normal surfaces. A *normal surface* is a planar surface that is orthogonal to the principal planes. If we look at the multiview projections of the prism, we see that surface A appears as an area in the top view, while in the

Figure 4-35 Rendered view of the complex part in Figure 4-34

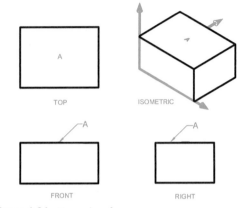

Figure 4-36 Normal surfaces

other views surface A appears as an edge. Also notice that because surface A is parallel to the horizontal projection plane, it is shown true size (TS) in the top view. In a three-view drawing a normal surface appears as a true-size surface in one view and on edge in the other two views.

INCLINED SURFACES

Surface B in Figure 4-37 is called an inclined surface. An *inclined surface* can be described as a normal surface that has been rotated about a line parallel to a principal axis. An inclined surface is perpendicular to one principal plane, and inclined (i.e., neither parallel nor perpendicular) to the others. An inclined surface appears as a

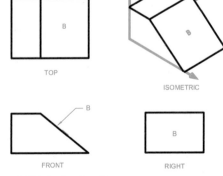

Figure 4-37 Inclined surfaces

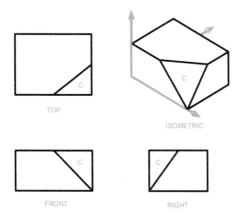

TOP

ISOMETRIC

FRONT

RIGHT

Figure 4-38 Oblique surfaces

foreshortened (i.e., not true size) surface in two views, while in the third view the inclined surface appears on edge. This edge length is a true length.

OBLIQUE SURFACES

An *oblique surface* is a planar surface that has been rotated about two principal axes. An oblique surface is inclined to all three principal projection planes. Surface C in Figure 4-38 is an oblique surface. Notice that surface C appears as an area in all three multiviews; in none of these views is the oblique surface true size.

Projection Studies

One way to improve visualization skills is to study four views (i.e., three multiviews, one pictorial) of simple objects like those appearing in Figure 4-39 on page 108. These projection studies improve the ability to recognize common shapes and features in combination.

Adjacent Areas

The top view in Figure 4-40 on page 109 has three distinct areas. Since no two adjacent areas can lie in the same plane, these areas must represent different surfaces. Some possible objects matching this view are also shown in the figure.

Surface Labeling

In order to help interpret a multiview drawing it is sometimes useful to label these surfaces, as shown in Figure 4-41 on page 109. Notice that surface 1 is a normal surface. It appears as an area in the top view and as an edge in the other views. Surface 3 is an inclined surface, appearing on edge in the front view and as a foreshortened area in the other views. Surface 5 is an oblique surface, and appears as a foreshortened area in all three views.

Similar Shapes

Unless viewed on edge, a planar face will always be projected with the same number of vertices. In addition, these vertices will always connect in the same sequence, no matter what the view. These facts are useful when reading a multiview drawing containing either inclined or oblique surfaces (see Figure 4-42 on page 109). Recall that an inclined surface appears as a foreshortened surface in two of three views, while an oblique appears as a foreshortened surface in all three views.

Vertex Labeling

In addition to labeling surfaces, it is also at times useful to label the vertices of a complicated surface in one view, and then to project these points into adjacent or related views. See Figure 4-43 on page 110 for an example, where the vertices of an oblique surface are labeled in one view, and then projected to the other views. Notice the similar shape of this oblique surface in the different views.

Analysis by Feature

As we will see in greater detail in Chapter 7, parts are built up from features. These features include such three-dimensional shapes as extrusions, revolutions, holes, ribs, chamfers, etc. By combining features we arrive at a completed part. See for example Figure 4-44 on page 110, where a part is built up from various features, including an extrusion, a boss (raised cylinder), a counterbore hole, a rib, and fillets and chamfers. Figure 4-45 on

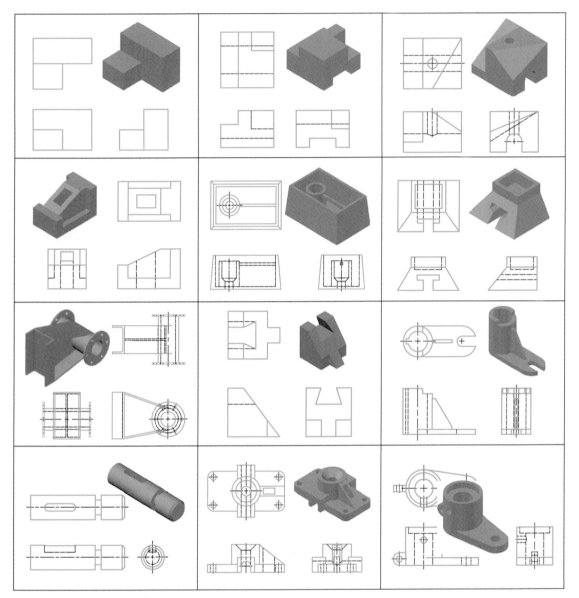

Figure 4-39 **Projection studies**

lize. If however, we break the part down into recognizable features, this task becomes more manageable. See Figure 4-47 on page 111 for a breakdown of the features of the object depicted in Figure 4-46.

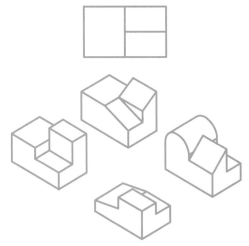

Figure 4-40 Adjacent areas

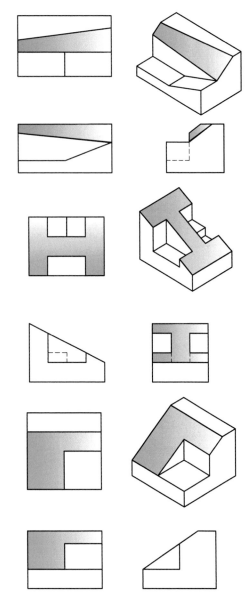

page 110 shows a multiview drawing of the part shown in Figure 4-44. Notice how these different manufacturing features appear in the multiview drawing. For example, a counterbore feature always appears as two concentric circles in the circular view, while in the rectangular view this feature appears as two rectangles stacked one on top of the other. Knowledge of how common manufacturing features appear in a multiview drawing can be used to advantage when called upon to interpret more complicated drawings.

Now look at the multiview drawing shown in Figure 4-46 on page 110. Without the benefit of a pictorial view, this object is difficult to visua-

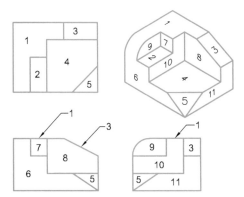

Figure 4-41 Surface labeling

Figure 4-42 Similar shapes

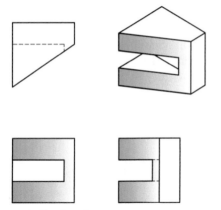

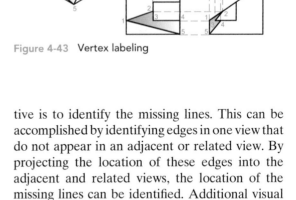

Figure 4-43   Vertex labeling

Figure 4-42   Continued.

Missing-Line and Missing-View Problems

Two additional tools, or rather exercises, which are very useful in developing spatial reasoning are missing-line and missing-view problems. In a missing-line problem three views are given, but some lines are missing from the views. The objec-

tive is to identify the missing lines. This can be accomplished by identifying edges in one view that do not appear in an adjacent or related view. By projecting the location of these edges into the adjacent and related views, the location of the missing lines can be identified. Additional visual reasoning is required to identify the type (e.g., visible, continuous) and extent of the missing lines. Figures 4-48 on page 111 and 4-49 on page 111 provide two examples of missing-line problems.

| A. Extrude | B. Boss (raised cylinder) | C. Counterbore hole | D. Chamfer (2) | E. Fillet (2) | F. Rib |

Figure 4-44   Breakdown of a part by features

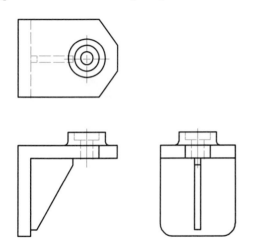

Figure 4-45   Multiview drawing of part

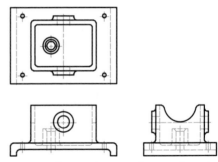

Figure 4-46   Multiview drawing of a complicated part

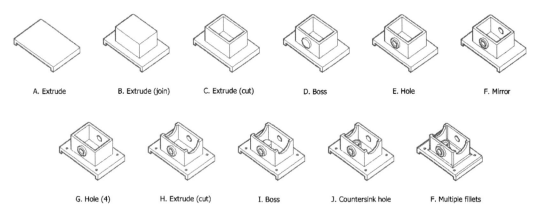

A. Extrude B. Extrude (join) C. Extrude (cut) D. Boss E. Hole F. Mirror

G. Hole (4) H. Extrude (cut) I. Boss J. Countersink hole F. Multiple fillets

Figure 4-47 Breakdown of a complicated part by features

More challenging still are missing-view problems. Here two of three views are given. The objective is to find the missing third view. As with the missing-line problems, edge features in the given views can be projected to help identify the lines in the missing view. This technique is employed, for example, in Figures 4-50 and 4-51 on page 112. In nearly all missing-view problems, however, it is even more helpful to sketch a well-proportioned pictorial view of the object. Start by sketching the object's bounding box. Next use the given views to identify prominent object features. The right side view of the object in Figure 4-51, for example, suggests a backwards "C" shape extrusion. Add this feature to the pictorial sketch. The vertical hidden line in the right side view still

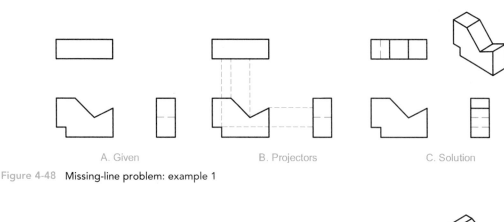

A. Given B. Projectors C. Solution

Figure 4-48 Missing-line problem: example 1

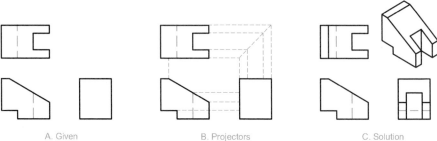

A. Given B. Projectors C. Solution

Figure 4-49 Missing-line problem: example 2

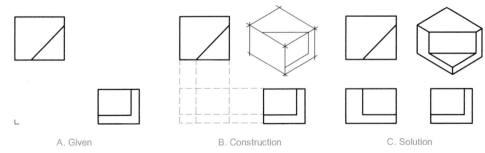

Figure 4-50 **Missing-view problem: example 1**

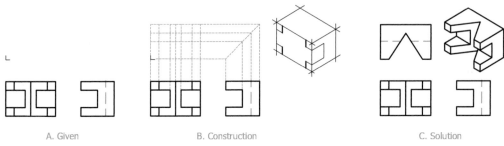

Figure 4-51 **Missing-view problem: example 2**

needs to be accounted for. By employing visual reasoning and some trial and error, it is found that a wedge-shaped vertical cut can account for this hidden line, as well as the internal vertical lines in the front view. The missing object is consequently composed of a "C" shape extrusion and the symmetrical wedge cut.

☐ QUESTIONS

TRUE AND FALSE

1. In a multiview drawing, it is always necessary to include at least three principal views in order completely to define the object.

2. In a multiview drawing, the right side view should be used, even if it has more hidden lines than the left view.

3. In a three-view drawing, an inclined planar surface will appear as a line in two of the principal views.

4. The angle at which the line-of-sight pierces the projection plane is the same for both multiview and axonometric projections.

MULTIPLE CHOICE

5. Which planar surface appears as a foreshortened surface in all of the standard multiviews?

   a. Normal

   b. Inclined

   c. Oblique

   d. Single curved

6. The glass box theory is used to describe:

   a. First angle projection of a part

   b. How multiviews are arranged with respect to one another

   c. How orthographic projections are made

   d. How isometric views are created

SKETCHING

7. Given the isometric view of the cut block objects appearing in P3-4 through P3–65, use Dwg 4-1 (or download worksheet from the book Web site) to sketch a multiview set (three views) of the object.

8. Given the isometric view of the cut block objects appearing in P3-4 though P3-65, use a blank sheet of paper to create a well-proportioned multiview sketch (three views) of the object.

9. Given the multiview set of the cut block objects appearing in P4-1 through P4-71, use Dwg 4-2 (or download worksheet from the book Web site) to sketch an isometric view of the object.

10. Given the multiview set of the cut block objects appearing in P4-1 through P4-71, use a blank sheet of paper to create a well-proportioned isometric sketch of the object.

11. Given the two views appearing in P4-72 through P4-102, use Dwg 4-1 and Dwg 4-2 (or download worksheets from the book Web site) to sketch the missing view and an isometric view of the object.

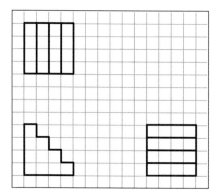

Figure P4-1

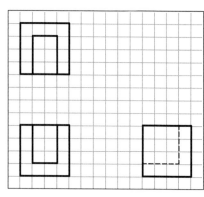

Figure P4-2

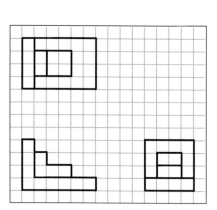

Figure P4-3

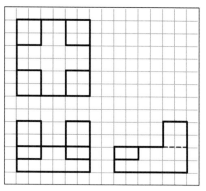

Figure P4-4

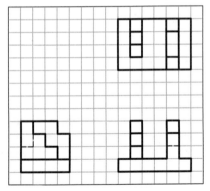

Figure P4-5

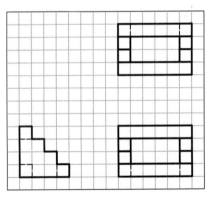

Figure P4-6

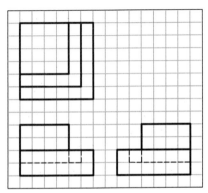

Figure P4-7

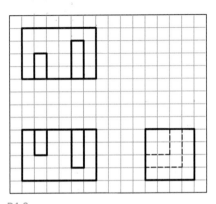

Figure P4-8

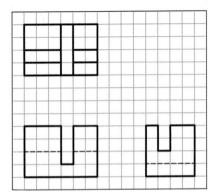

Figure P4-9

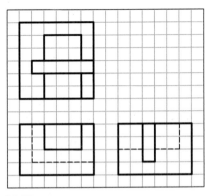

Figure P4-10

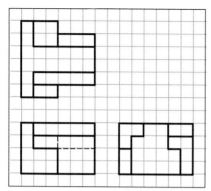

Figure P4-11

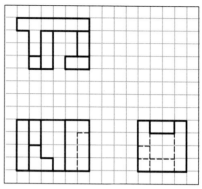

Figure P4-12

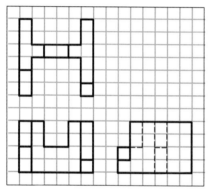

Figure P4-13

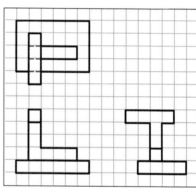

Figure P4-14

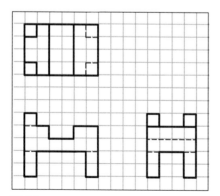

Figure P4-15

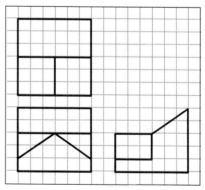

Figure P4-16

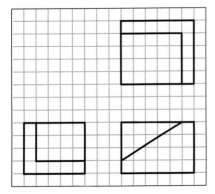

Figure P4-17

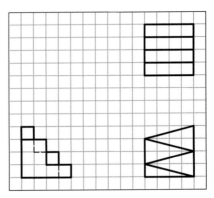

Figure P4-18

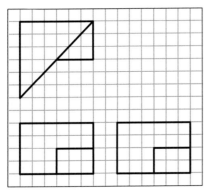

Figure P4-19

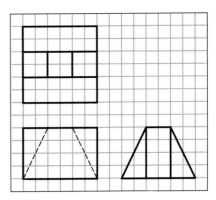

Figure P4-20

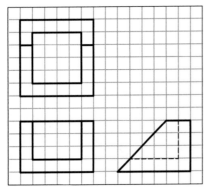

Figure P4-21

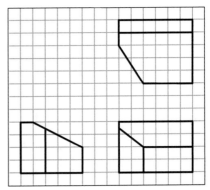

Figure P4-22

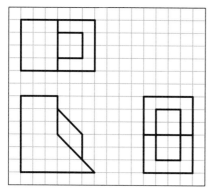

Figure P4-23

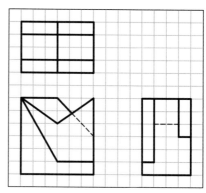

Figure P4-24

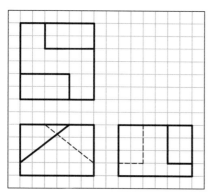

Figure P4-25

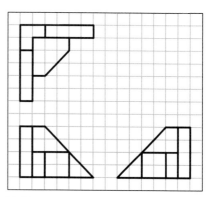

Figure P4-26

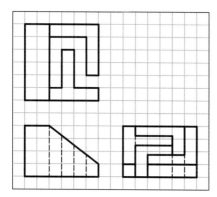

Figure P4-27

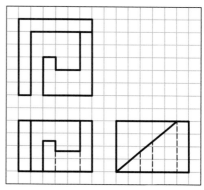

Figure P4-28

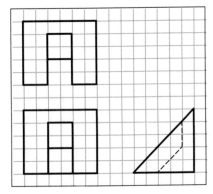

Figure P4-29

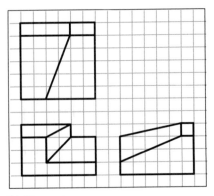

Figure P4-30

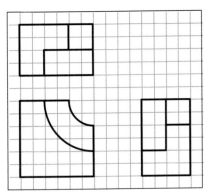

Figure P4-31

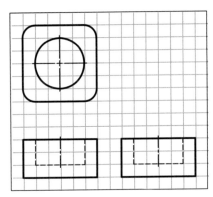

Figure P4-32

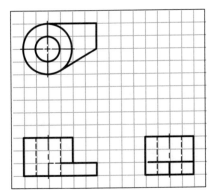

Figure P4-33

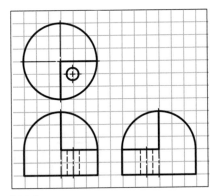

Figure P4-34

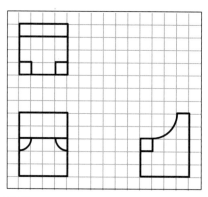

Figure P4-35

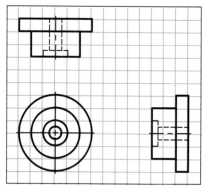

Figure P4-36

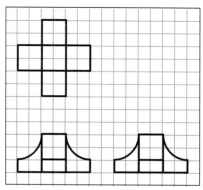

Figure P4-37

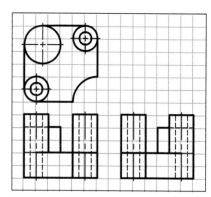

Figure P4-38

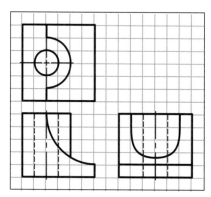

Figure P4-39

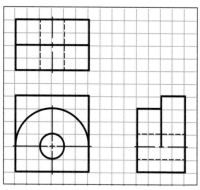

Figure P4-40

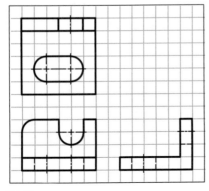

Figure P4-41

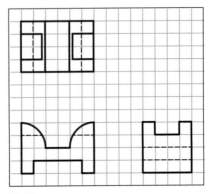

Figure P4-42

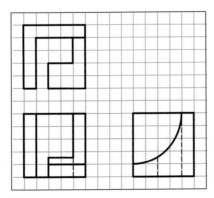

Figure P4-43

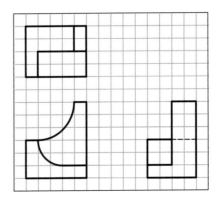

Figure P4-44

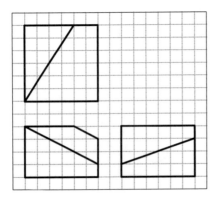

Figure P4-45

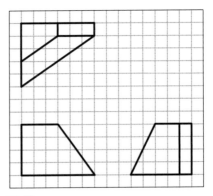

Figure P4-46

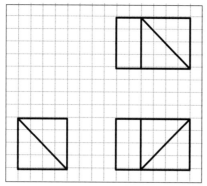

Figure P4-47

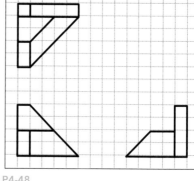

Figure P4-48

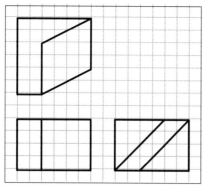

Figure P4-49

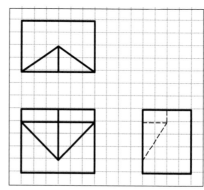

Figure P4-50

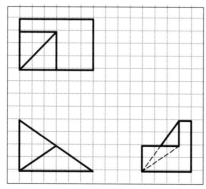

Figure P4-51

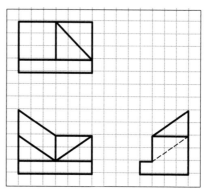

Figure P4-52

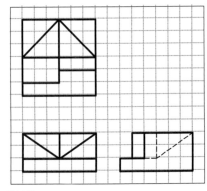

Figure P4-53

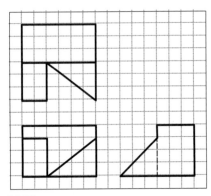

Figure P4-54

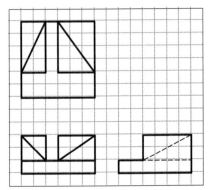

Figure P4-55

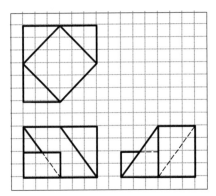

Figure P4-56

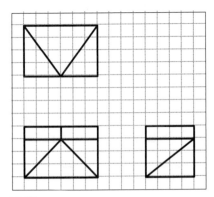

Figure P4-57

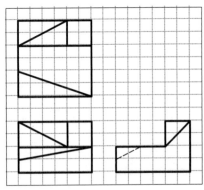

Figure P4-58

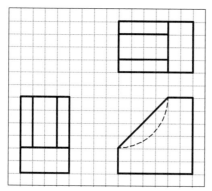

Figure P4-59

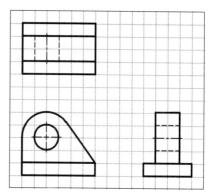

Figure P4-60

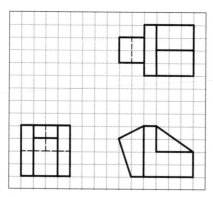

Figure P4-61

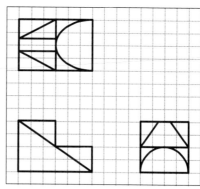

Figure P4-62

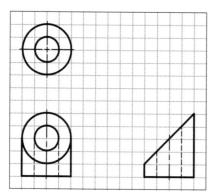

Figure P4-63

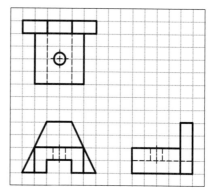

Figure P4-64

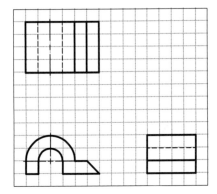

Figure P4-65

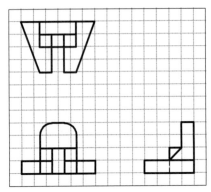

Figure P4-66

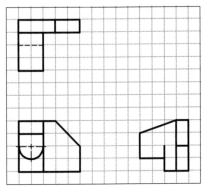

Figure P4-67

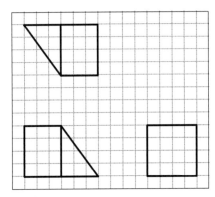

Figure P4-68

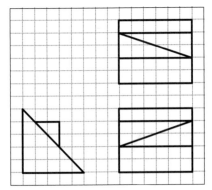

Figure P4-69

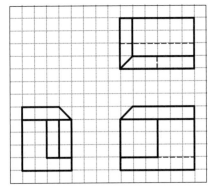

Figure P4-70

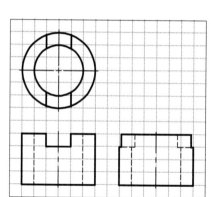
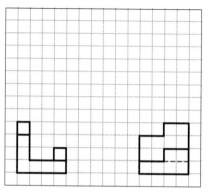

Figure P4-71

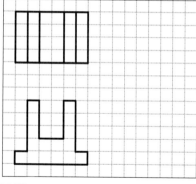

Figure P4-72

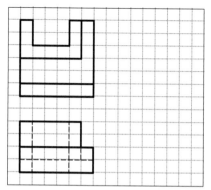

Figure P4-73

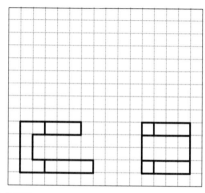

Figure P4-74

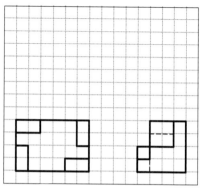

Figure P4-75

Figure P4-76

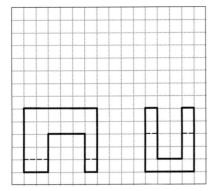

Figure P4-77

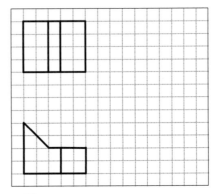

Figure P4-78

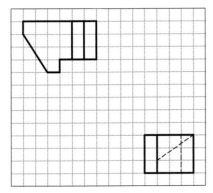

Figure P4-79

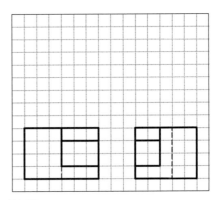

Figure P4-80

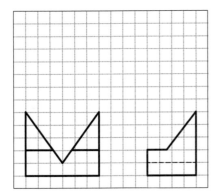

Figure P4-81

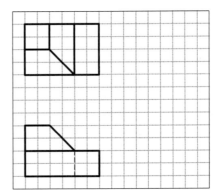

Figure P4-82

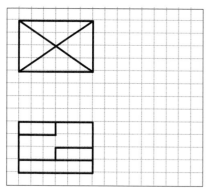

Figure P4-83

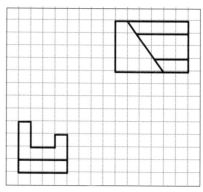

Figure P4-84

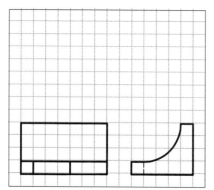

Figure P4-85

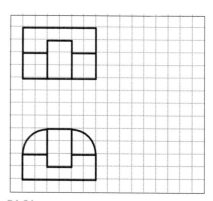

Figure P4-86

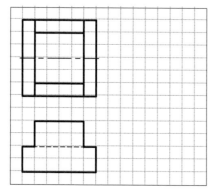

Figure P4-87

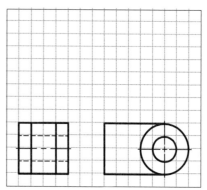

Figure P4-88

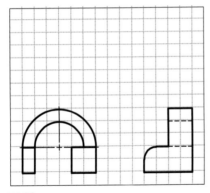

Figure P4-89

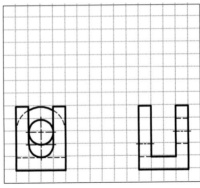

Figure P4-90

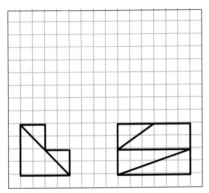

Figure P4-91

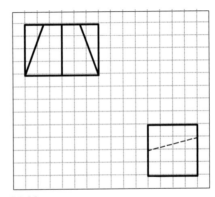

Figure P4-92

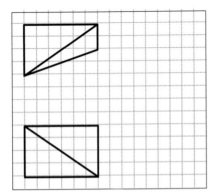

Figure P4-93

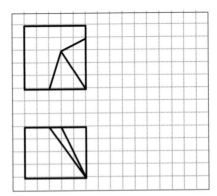

Figure P4-94

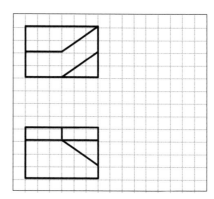

Figure P4-95

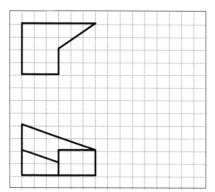

Figure P4-96

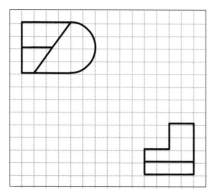

Figure P4-97

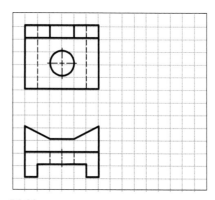

Figure P4-98

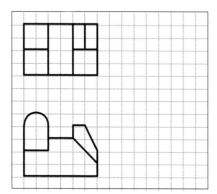

Figure P4-99

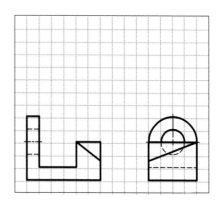

Figure P4-100

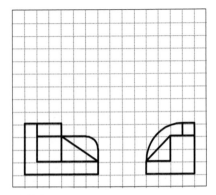

Figure P4-101

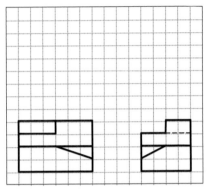

Figure P4-102

5 AUXILIARY AND SECTION VIEWS

☐ AUXILIARY VIEWS

Introduction

Recall that in a multiview drawing of an object with an inclined surface, in one multiview the inclined surface is seen on edge, while in the other two multiviews the inclined surface appears as a foreshortened (i.e., not true size) surface. In Figure 5-1, on page 132 for example, the inclined surface labeled A is seen on edge in the top view, and as a foreshortened surface in the front and right side views. In many circumstances the true size of an inclined surface is very useful.

From ***descriptive geometry***[1] it is known that the true size and shape of a planar face (or the true length of a line) can only be represented in an orthographic projection if the line of sight is normal to the planar face or, equivalently, if the projection plane is parallel to the face. This knowledge will be put to use to find the true size of an inclined surface.

Definitions

In earlier chapters we saw that multiview projection is an orthographic projection technique by which a three-dimensional object is projected onto one of three mutually perpendicular planes. These are the principal planes; horizontal, frontal, and profile. An ***auxiliary view*** is an orthographic view that is projected onto any plane

other than one of the principal planes. A ***primary auxiliary view*** is an auxiliary view that is projected onto a plane perpendicular to one of the principal planes, and inclined to the other two. A primary auxiliary view can be used to find the true size and shape of an inclined surface. A ***secondary auxiliary view*** is projected from a primary auxiliary view onto a plane that is inclined to all three principal projection planes. A secondary auxiliary view can be used to find the true size and shape of an oblique surface.

Auxiliary View Projection Theory

Figure 5-2 on page 132 shows an object with an inclined surface that has been placed inside a glass box. Note that the glass box has been modified by adding an additional plane that is parallel to the inclined surface. For the situation shown in Figure 5-2 the inclined plane (i.e., auxiliary plane) is perpendicular to the frontal plane and inclined to the horizontal and profile planes.

In Figure 5-3 on page 132 orthographic projection is used to project the object onto the projection planes (i.e., sides of the glass box), including the auxiliary plane. Since the auxiliary plane is parallel to the inclined surface, the resulting projection shows the true size and shape of the inclined surface, as well as foreshortened projections of any other visible surfaces. Also notice that the edge view of the inclined surface appears on the frontal projection plane.

Now imagine that the frontal plane is hinged to the horizontal, profile, and auxiliary planes. If the hinged views are then unfolded so that they lie in the same plane as the frontal view, Figure 5-4 on page 132 results. Note that the distance D from

---

[1] The term descriptive geometry refers to the body of knowledge, developed over the centuries, consisting of mathematical-graphical procedures used to accurately describe 3D geometry within 2D media. The French mathematician Gaspard Monge (1746–1818) is considered the father of descriptive geometry.

132 □ Chapter 5 AUXILIARY AND SECTION VIEWS

132 □ Chapter 5 AUXILIARY AND SECTION VIEWS

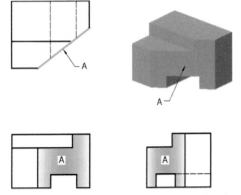

Figure 5-1 Views of cut block with an inclined surface

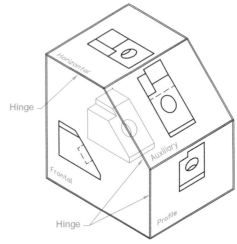

Figure 5-3 Views projected onto sides of glass box

the hinge line to the near side of the inclined surface is the same for all three projected views (i.e., horizontal, profile, and auxiliary). This fact will be used later in the construction of an auxiliary view sketch of an inclined surface.

As seen in Figure 5-5 on the following page, an auxiliary view is in alignment with the principal view that shows the inclined surface on edge. Since these inclined edges are projected true length, the perpendicular distances between the dashed lines shown in Figure 5-5 represent the actual inclined edge lengths on the inclined surface.

Note that the auxiliary view in Figure 5-5 only shows the inclined surface. This is called a *partial auxiliary view*. Since they are both easier to execute and to visualize than full auxiliary views, partial auxiliary views are frequently employed. In creating an auxiliary projection with a CAD system, however, a full auxiliary view is obtained automatically. The resulting view can always be modified by either hiding or deleting lines in order to obtain a view that only shows the inclined surface of interest.

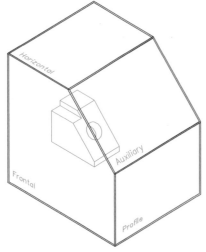

Figure 5-2 Glass box modified for auxiliary view

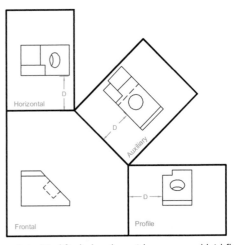

Figure 5-4 Modified glass box sides open and laid flat

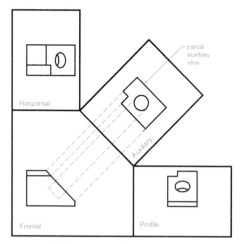

Figure 5-5 Auxiliary view alignment

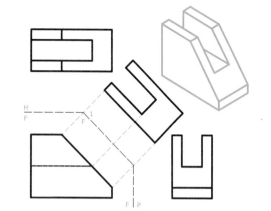

Figure 5-7 Auxiliary view projected from frontal plane

Auxiliary Views: Three Cases

The primary auxiliary view is projected from the principal plane containing the edge view of the inclined surface. If an object with an inclined surface is oriented as shown in Figure 5-6, then the edge view of the inclined surface appears on the horizontal projection plane. In this case, the auxiliary view is projected from the horizontal plane. Notice how the hinge lines between the horizontal and frontal (H–F) and the horizontal and auxiliary (H–1) planes are represented.

With the same object oriented as shown in Figure 5-7, the inclined surface appears on

edge in the frontal plane. In this case the primary auxiliary view is projected from the frontal plane.

In the third case, shown in Figure 5-8, the edge view of the inclined surface appears in the profile plane. The primary auxiliary view is consequently projected from the profile plane.

General Sketching Procedure for Finding a Primary Auxiliary View

In this section the sketching procedure for obtaining a primary auxiliary view is outlined. The problem can be stated as follows: given a multi-view drawing of an object with an inclined surface

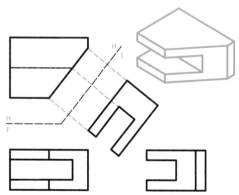

Figure 5-6 Auxiliary view projected from horizontal plane

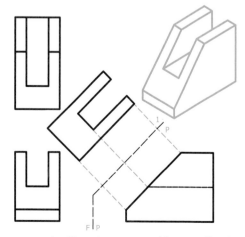

Figure 5-8 Auxiliary view projected from profile plane

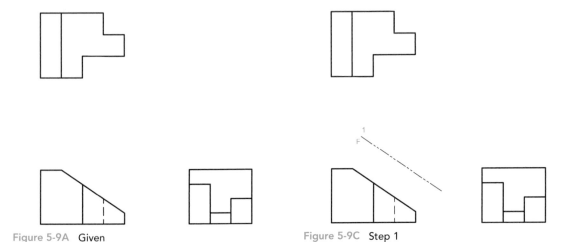

Figure 5-9A Given

Figure 5-9C Step 1

(see Figure 5-9A), find the primary auxiliary view showing the true size and shape of this inclined surface (see Figure 5-9B). The process takes advantage of the fact that the inclined surface is shown as a true length edge in one of the multiviews. Perpendicular projectors are erected from the edge view to obtain these distances. All that is required to complete a partial auxiliary view are the edge lengths perpendicular to the inclined edge. These distances are available in the views adjacent to the view containing the inclined surface on edge. A trammel can be used to capture and then transfer these edge lengths to the auxiliary view. Two reference edges, one for the adjacent/related view, the other for the auxiliary

view, assist in the transferring of the distances. These reference edges are comparable to the glass box hinge lines shown in Figures 5-3 and 5-4, but are more flexible. In the solution shown in Figure 5-9B, for example, the reference edge labeled H–F is conveniently located so that it is colinear with the near side edge of the inclined surface, as shown in the top view. This is equivalent to setting $D = 0$ in Figure 5-4, and reduces the number of required trammel dimensions by one.

STEP 1 Sketch a dashed reference line parallel to the edge view of the inclined surface to be projected (see Figure 5-9C).

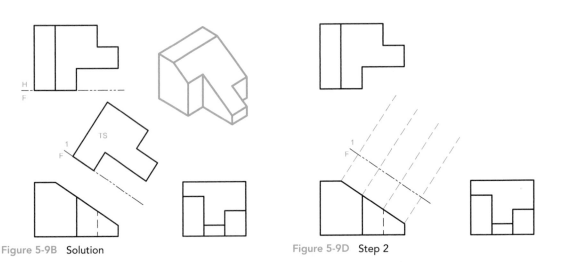

Figure 5-9B Solution

Figure 5-9D Step 2

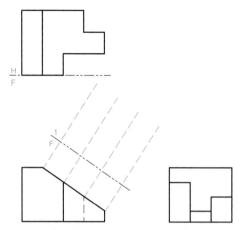

Figure 5-9E **Step 3**

or P) indicates the adjacent or related view from which the missing dimensional information is to be obtained.

- In Figure 5-9E, reference edge F–H will be used to capture depth information from the top view. Alternatively, a reference edge F–P could have been used to capture this same depth information from the profile view.

- This second reference line represents a hinge, where the adjacent/related view is rotated into the same plane as the other views.

STEP 4 (OPTIONAL) In this optional step, label the vertices of the inclined surface in the adjacent/related view, then transfer the vertex labels to the edge view (see Figure 5-9F).

- This step may be helpful in correctly orienting the auxiliary view.

STEP 5 Using a trammel or dividers, transfer the depth dimensions from F–H reference line in the top view to the F–1 reference line (see Figure 5-9G).

- The vertex labels can also be transferred to the auxiliary view.

STEP 6 Knowing the placement of the edges, the inclined surface can be sketched. It is customary to label the projected surface with "TS," indicating that it is true size (see Figure 5-9H).

- The perpendicular distance between the reference line and the inclined edge should be chosen so that the resulting auxiliary view does not interfere with other views.

- This reference line is labeled either H–1, F–1, or P–1, depending on whether the edge view of the inclined surface appears in the horizontal (H), frontal (F), or profile (P) plane. The "1" indicates that a primary auxiliary view is being constructed.

- The reference line represents the hinge that the auxiliary view is rotated 90 degrees about, thus causing it to lie in the same plane as the principal plane it is projected from.

STEP 2 Sketch perpendicular projectors from the inclined surface (see Figure 5-9D on the previous page).

- The edge view of the inclined surface shows the true lengths of these edges, since these edges are parallel to the projection plane.

- The order of steps 1 and 2 are interchangeable.

STEP 3 Draw a second dashed reference line at a convenient location between the view being projected from and an adjacent or related principal view (see Figure 5-9E).

- Label the second reference line so that the first letter indicates the view being projected from (either H, F, or P), and the second letter (H, F,

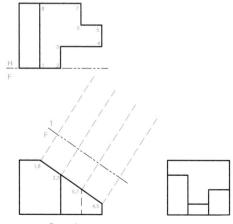

Figure 5-9F **Step 4**

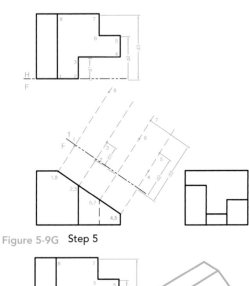

Figure 5-9G **Step 5**

Figure 5-9H **Step 6**

Finding a Primary Auxiliary View of a Contoured Surface

In the example shown in Figure 5-10, the procedure for finding an auxiliary view is repeated. However, this time the inclined surface has a curved profile. Note that the partial symmetry of this surface is exploited by choosing the H–1 reference edge to lie along the axis of symmetry.

☐ SECTION VIEWS

Introduction

Parts like the one shown in Figure 5-11 on the following page contain several internal features. These interior construction details show up as hidden lines in a standard multiview projection. Since hidden lines can be difficult to interpret and visualize, **section views** are frequently used to expose the internal features of a part.

Section View Process

In a section view, an imaginary **cutting plane** is passed through the part, often along a part's plane of symmetry. The portion of the part between the viewer and the plane is removed, and the part is

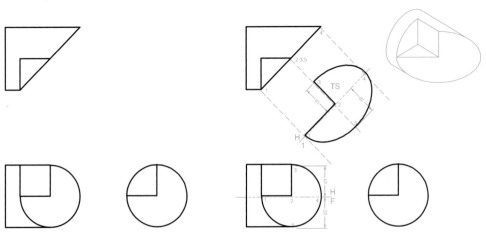

<table>
<tr><td>Given</td><td>Solution</td></tr>
</table>

Figure 5-10 Auxiliary view process applied to a contoured, symmetrical surface

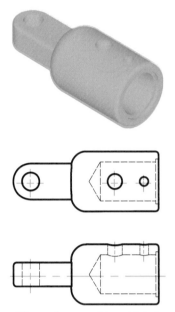

Figure 5-11   Views of a part with multiple internal features

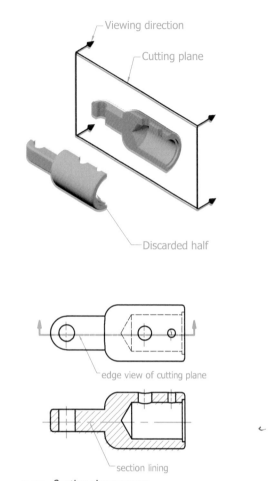

Figure 5-12   Section view process

then viewed perpendicular to the cutting plane (see Figure 5-12). A *section lining*, or hatch pattern, is applied to the surfaces that make contact with the cutting plane.

The cutting plane seen on edge shows the location of the section. This edge view of the cutting plane is shown in the view from which the section is projected. The cutting plane edge is typically represented as a thick dashed line. Note that a cutting plane line has precedence over a centerline, should the two coincide. The viewing direction is indicated by arrows drawn perpendicular to the cutting plane.

Lines that would be visible after making a cut are also shown in a section view; see Figure 5-13 on page 138. By convention, hidden lines are normally not shown in a section view. Exceptions are occasionally made, however, if it is felt that the clarity of the drawing is improved.

Section views are typically labeled using indexed capital letters, for example Section A–A, Section B–B, etc. The capital letters are also used to label the cutting plane line, and to clearly associate the two views, as shown in Figure 5-14 on page 138.

Section Lining (Hatch Patterns)

As we have already seen, section lining is applied to solid areas on the part that have been exposed by the cutting plane. As shown in Figure 5-15 on page 138, section line types are often associated with different materials. The most commonly employed section lining consists of uniformly spaced continuous lines, set at a 45-degree angle. This section lining angle, however, should be adjusted to avoid the section lines being too close to parallel or perpendicular with the visible lines that bound them, as seen in Figure 5-16 on page 139. In CAD software programs, section lines are typically referred to as hatch patterns.

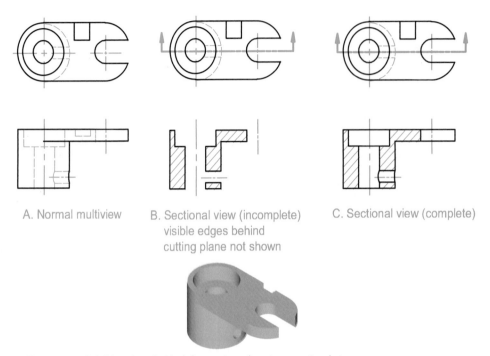

A. Normal multiview

B. Sectional view (incomplete)
visible edges behind
cutting plane not shown

C. Sectional view (complete)

Figure 5-13   Treatment of visible edges behind the cutting plane in a sectional view

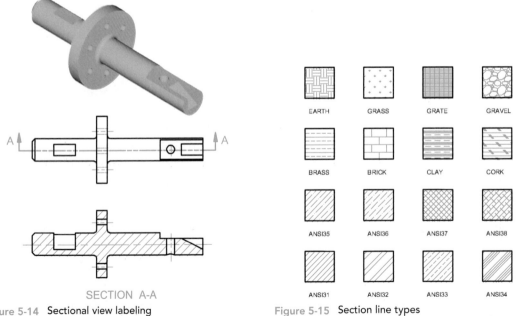

SECTION A-A

Figure 5-14   Sectional view labeling

Figure 5-15   Section line types

EARTH   GRASS   GRATE   GRAVEL

BRASS   BRICK   CLAY   CORK

ANSI35   ANSI36   ANSI37   ANSI38

ANSI31   ANSI32   ANSI33   ANSI34

Figure 5-16   Adjusting the section lining angle

A. Avoid   B. Preferred

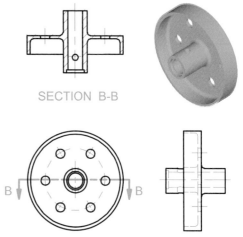

SECTION B-B

Full Sections

In a *full section*, the cutting plane passes all the way through the object. Figure 5-17 shows another example of a full section view that appears in the front view.

Another possibility, typically one more difficult to visualize, uses a horizontally oriented cutting plane. In this situation (see Figure 5-18), the cutting plane appears on edge in the front

Figure 5-18   Full section appearing in the top view

view, while the section appears in the top view. Notice the direction of the cutting plane arrows in the front view. The arrow direction represents the viewing direction; the observer is looking down at the bottom portion of the object, the top portion having been removed.

Yet another possible orientation of a full section view appears in Figure 5-19. Here the section appears in the (left) side view, with the cutting plane edge appearing in the front view. Once again, note the direction of the sight arrows.

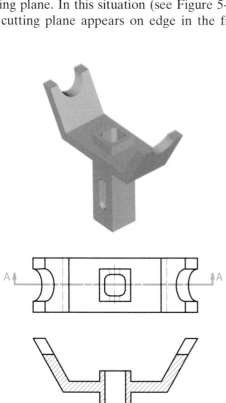

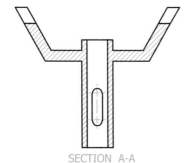

SECTION A-A

Figure 5-17   Full section appearing in the front view

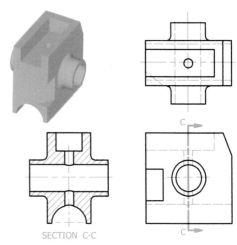

SECTION C-C

Figure 5-19   Full section appearing in the side view

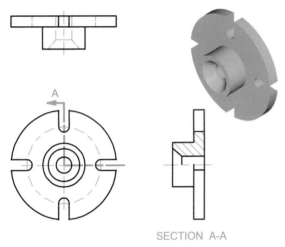

SECTION A-A

Figure 5-20   Half section view

A. Incorrect

B. Correct

Also note that the cutting edge could just as well have appeared in the top view.

Half Sections

With symmetrical or very nearly symmetrical objects, it is not always necessary to pass the cutting plane all the way through the part. In a *half section*, the cutting plane only passes half way through the part, as shown in Figure 5-20. In a half section view, one quarter of the part is removed.

Half sections possess the advantage of showing both the interior and the exterior of the part in the same view. External features are included on the unsectioned half. A centerline is used to separate the two halves. Hidden lines are normally omitted in both halves, but may be shown in the un-sectioned half.

Offset Sections

An *offset section* is a modified full section that is used when important internal features do not lie in the same plane. In an offset section the cutting plane is stepped, or offset in order to pass through these features. Figure 5-21 shows an example of an offset section. Notice that any steps (90-degree bends) in the cutting plane line are not shown in

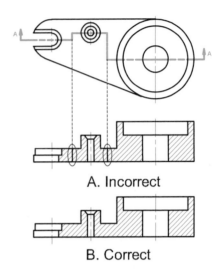

Figure 5-21   Offset section view

the section view. The section is drawn as if these offsets all lie in the same plane. Also notice that the steps should be located in regions where there are no features.

Broken-out Sections

A *broken out section* is used when only a portion of the part needs to be sectioned. See Figure 5-22 on page 141 for an example of a broken-out section. A jagged or freehand break line is used to separate the sectioned from the unsectioned portion of the drawing. Like half sections, broken-out sections have the advantage of showing internal and external features in the same view. In addition to being used on multiview drawings, broken-out sections are also used on pictorial views, particularly when executed in CAD (see Figure 5-23 on page 141).

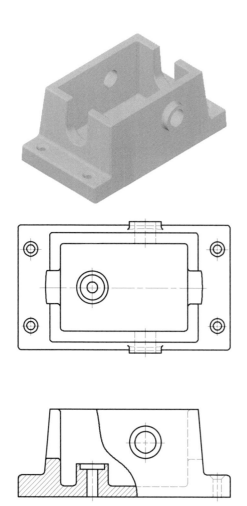

Figure 5-22 Broken out section view

Revolved Sections

In all of the sections discussed thus far (full, half, offset, and broken-out), the section view is pro-

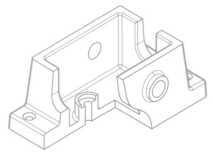

Figure 5-23 Broken-out section view—pictorial

jected from the adjacent view in which the cutting plane line appears. A **revolved section**, on the other hand, is created by passing a cutting plane perpendicular to the center place of an elongated symmetrical feature, and then revolving the resulting cross-section 90 degrees into the plane of the drawing. This results in the cross-section being superimposed on the original view. Figure 5-24 on page 142 shows an example of a revolved section. The original section may be shown with (Figure 5-24A) or without conventional break lines (Figure 5-24B). A centerline is used to represent the axis of the revolved section.

Removed Sections

A **removed section** is similar to a revolved section, except that the cross-section is not superimposed on the view. Rather, the removed section is placed at some convenient location. Standard section view labeling practices are used to relate the cutting plane location to the resulting section. See Figure 5-25 on page 142 for an example of a removed section. Removed sections are used when there is insufficient room for a revolved section, and when several cross-sections are needed to show the transition of an elongated feature from one shape to another (see Figure 5-26 on page 143).

Conventional Representations: Section Views

In order to simplify the construction and improve the clarity of section views, conventional representations are sometimes employed in place of true orthographic projections. These simplified representations are widely recognized and accepted as being a part of standard drawing practice. Conventional representations associated with section views include the treatment of thin features, radially distributed and off-angle features, and section lining in assembly sections. Note that when using CAD it may actually be easier to obtain a true projection, rather than a simplified representation. Although this has,

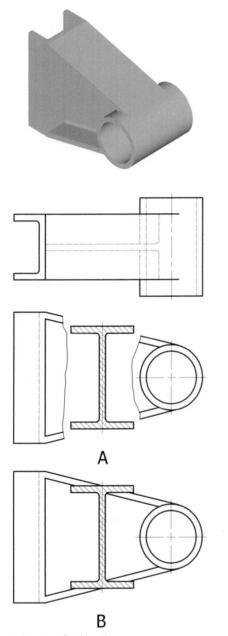

A

B

Figure 5-24   Revolved section

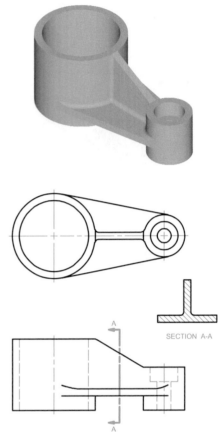

SECTION A-A

Figure 5-25   Removed section

Conventional Representations: Thin Features

In an effort to make some section views more readable, section lining is not applied to the outline of thin features like ribs, webs, and lugs, when the cutting plane passes along the length of the feature. Figure 5-27 on page 143 provides an example of this convention applied to a rib feature.

Note that, without this convention, the section shown in Figure 5-28 on page 144 could be incorrectly interpreted as depicting a part with uniform thickness (Figure 5-28A), rather than as a ribbed part (Figure 5-28B).

Figure 5-29 on page 144 provides an example of this thin feature convention applied to both a lug and a web feature.

to some extent, reduced the usage of certain conventional representations, it is still necessary for engineers to be familiar with this aspect of the language of engineering graphics.

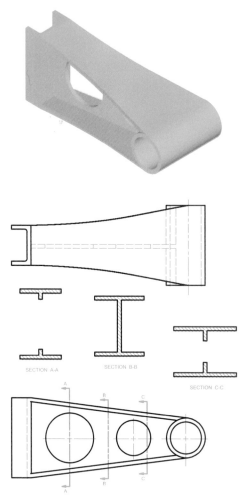

SECTION A-A

SECTION B-B

SECTION C-C

Figure 5-26 Multiple removed sections

Section View Construction Process

Figure 5-30 on page 145 illustrates the process of constructing a full section view of a part. In this particular problem only the top and right side principal views are given, and we are asked to draw the cutting plane on edge and to find the front section view. In a less difficult variant of this problem, a pictorial view of the object is given, along with the cutting plane location.

In Step 1, the cutting plane is drawn in the top view along the object's axis of symmetry. The arrows should point as shown, to indicate that the section will appear in the front view. Parallel

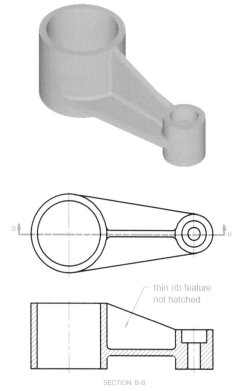

thin rib feature
not hatched

SECTION B-B

Figure 5-27 Conventional treatment of a thin rib feature in a sectional view

construction lines, or projectors, are drawn from the given views to help locate the extents of the object features in the front view.

Without the benefit of a pictorial view, visual reasoning must now be employed, reading between the given top and side views to help piece together a mental image of the object. In the top view on the left there are three concentric circles. A raised cylinder with a counterdrilled, counterbore, or countersunk hole feature will all appear like this in their circular view. Looking now at the right side view, it appears that this is a counterdrilled hole feature. Reading between the two given views, it also appears that a horizontally oriented hole is drilled through the raised cylinder, intersecting with the vertically oriented counterdrill feature. Note that these two holes account for all of the hidden lines appearing in the two given views.

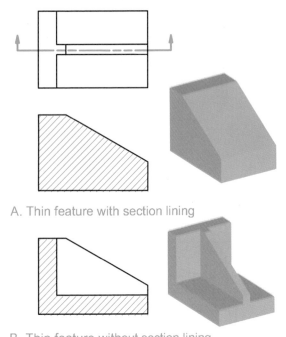

A. Thin feature with section lining

B. Thin feature without section lining

Figure 5-28 Justification of thin feature convention

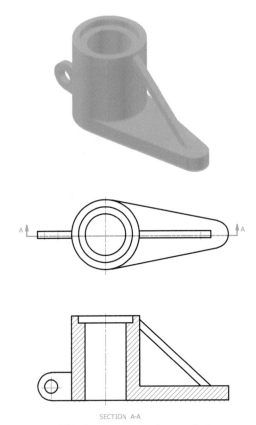

SECTION A-A

Figure 5-29 Thin feature convention applied to a part with a lug and web

Moving now to the right side of the object, as seen in the top view on the right, we see a feature tangent to the raised cylinder on the left, with a fillet and a semi-circular cut on the right. But what are the vertical extents of this feature? Looking at the right view we see three horizontally aligned rectangles, a little way up from the bottom. Together these views suggest a plate, rounded at one end with a semicircular cutout, attached to the raised cylinder.

At this point (Step 2) these features can be laid out. The only remaining feature, represented by the rectangular shapes close to the centerlines in the given views, top and right, suggests a support rib. Note that, based on the information provided in the given views, the rib profile could either have been straight-sided (as shown in Figure 5-29), or contoured (curved).

In Step 3 section lining is applied to the solid areas through which the imaginary cutting plane passes, except for the rib feature, which is left

without hatching, by convention. Note that the section lining angle has been adjusted from the default 45 to 30 degrees, as shown in Figure 5-16 on page 139, because of the inclined lines describing the counterdrill feature. Centerlines have also been added in Step 3.

Conventional Representations: Aligned Sections

This convention is used to simplify the construction of section views containing radially distributed and off angle features like holes, ribs, lugs, etc. In the object shown in Figure 5-31 on page 146, for example, a true orthographic projection (see Figure 5-31A) results in a difficult to interpret, foreshortened view of the rib on the left side. To eliminate these problems, by convention this

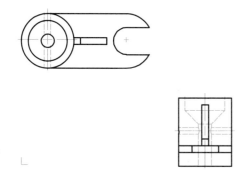

Figure 5-30A Full section view construction process: given

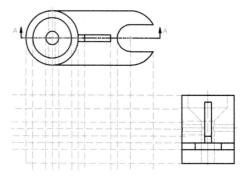

Figure 5-30B Step 1

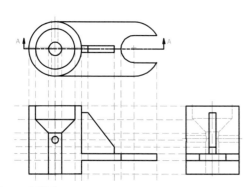

Figure 5-30C Step 2

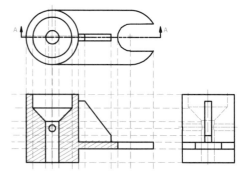

Figure 5-30D Step 3

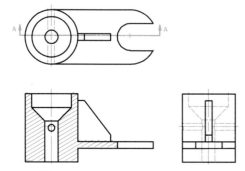

Figure 5-30E Completed sketch

Figure 5-30F Shaded view

feature is rotated into the cutting plane, or alternatively, the cutting plane is bent to pass through the feature. Similarly, the lug and hole features shown on the right side in Figure 5-31A are rotated into the cutting plane. The result is a clearer representation of the geometry, as shown in Figure 5-31B.

Another example of features being rotated or aligned to simplify the representation is shown in Figure 5-32.

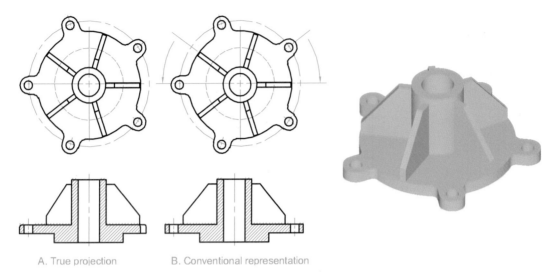

A. True projection   B. Conventional representation

Figure 5-31   Conventional representation of radially distributed features

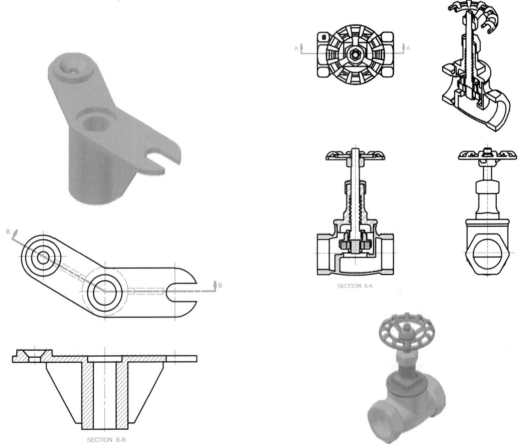

SECTION B-B

SECTION A-A

Figure 5-32   Conventional representation of an off-angle feature

Figure 5-33   Assembly section view (Courtesy of Alexander H. Hays)

Assembly Section Views

In a section view of an assembly, different hatch patterns are applied to different parts. See, for example, Figure 5-33, where several different section linings are used to represent different parts. Notice that thin-walled parts (e.g., shafts, nuts) are not sectioned. In addition, parts like washers, bushings, gaskets, bolts, screws, keys, rivets, pins, bearings, spokes, and gear teeth are not sectioned where the cutting plane lies along the longitudinal axis of the part.

□ QUESTIONS

TRUE AND FALSE

1. Primary auxiliary views are used to find the true size and shape of oblique surfaces.
2. Only the inclined face of an object projected onto an auxiliary plane is shown in a partial auxiliary view.
3. When a cutting plane is passed through the length of a thin feature such as a rib, section lining is not applied.

MULTIPLE CHOICE

4. The projection plane for a primary auxiliary view is:
   a. Parallel to one of the principal projection planes and perpendicular to the other two
   b. Perpendicular to one of the principal projection planes and inclined to the other one
   c. Perpendicular to two of the principal projection planes and inclined to the other one
   d. Inclined to all three of the principal projection planes
5. Which of the following is not a legitimate type of section view?
   a. Full
   b. Half

c. Quarter
d. Removed
e. Revolved
f. Aligned
g. Offset
h. Broken-out

SKETCHING AND MODELING

6. Given the two views appearing in P5-1 through P5-26, use Dwg5-1 (or download worksheet from the book Web site, www.wiley.com/college/leake) to sketch the auxiliary view of the object.
7. Given the two views appearing in P5-1 through P5-26, use Dwg5-1 and Dwg5-2 (or download worksheets from the book Web site) to sketch the auxiliary, missing, and isometric view of the object.
8. Given the dimensioned isometric view (P7-8, 10, 11, 13, 14), create a well-proportioned multiview sketch (three views) of the object, and find the auxiliary view of the inclined surface.
9. Given the dimensioned isometric view (P7-8, 10, 11, 13, 14), create a solid model of the object and generate a multiview drawing along with an auxiliary of the inclined surface.
10. Given the two views appearing in P5-27 through P5-32, use Dwg5-1 (or download worksheet from the book Web site) to sketch the section view of the object.
11. Given the two views appearing in P5-33 through P5-45, use Dwg5-1 (or download worksheet from the book Web site) to sketch the section view of the object with the cutting plane shown in the appropriate view.
    a. Full section (P5-33 through P5-37)
    b. Half section (P5-38 through P5-41)
    c. Offset section (P5-42 through P5-45)

12. Given the two views appearing in P5-46 through P5-60, use Dwg5-1 (or download worksheet from the book Website) to sketch the section view of the object with the cutting plane shown in the appropriate view.
    a. Full section (P5-46 through P5-48)
    b. Half section (P5-49 through P5-50)
    c. Offset section (P5-51 through P5-52)
    d. Broken-out section (P5-53 through P5-54)
    e. Removed section (P5-55 through P5-57)
    f. Revolved section (P5-55 through P5-57)
    g. Aligned section (P5-58 through P5-60)

13. Given the dimensioned isometric view (P7-2, 5, 6, 9, 13, 17), create a well-proportioned multiview sketch of the object and find a section view with the cutting plane shown in the appropriate view.

14. Given the dimensioned isometric view (P7-2, 5, 6, 9, 13, 17), create a solid model of the object and generate a multiview drawing with a section view.

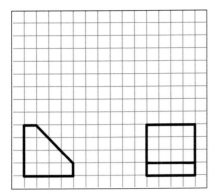

Figure P5-1

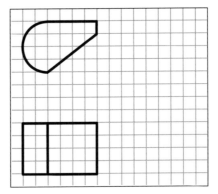

Figure P5-2

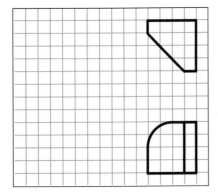

Figure P5-3

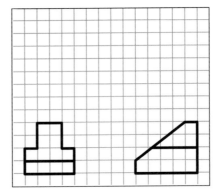

Figure P5-4

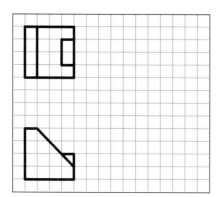

Figure P5-5

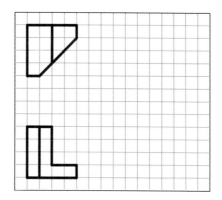

Figure P5-6

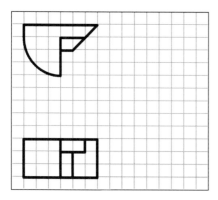

Figure P5-7

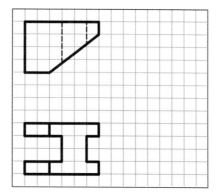

Figure P5-8

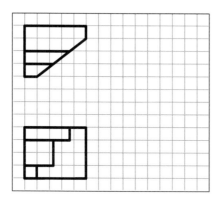

Figure P5-9

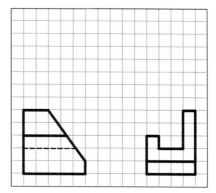

Figure P5-10

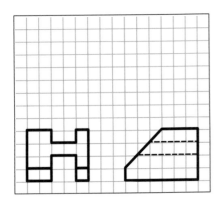

Figure P5-11

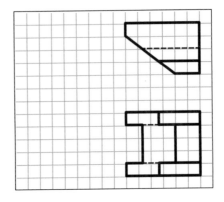

Figure P5-12

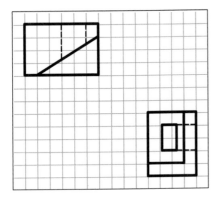

Figure P5-13

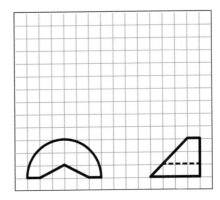

Figure P5-14

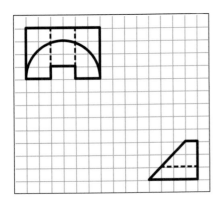

Figure P5-15

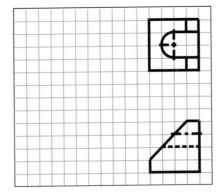

Figure P5-16

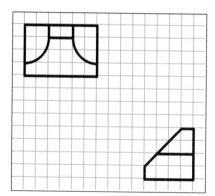

Figure P5-17

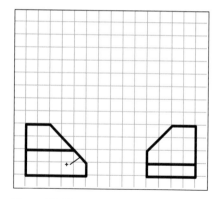

Figure P5-18

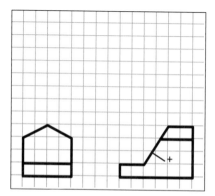

Figure P5-19

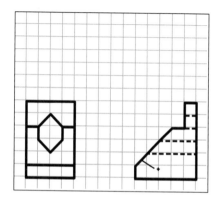

Figure P5-20

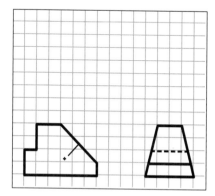

Figure P5-21

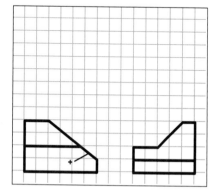

Figure P5-22

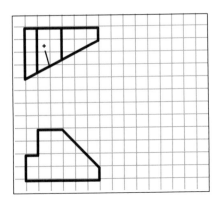

Figure P5-23

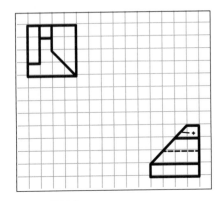

Figure P5-24

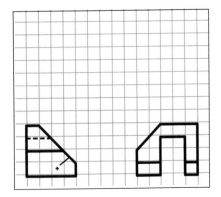

Figure P5-25

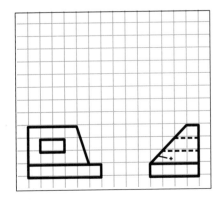

Figure P5-26

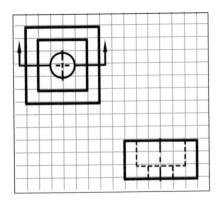

Figure P5-27

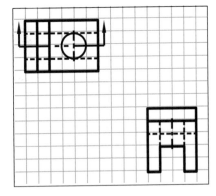

Figure P5-28

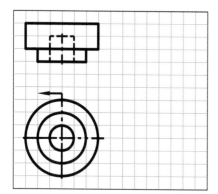

Figure P5-29

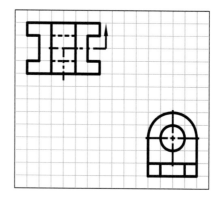

Figure P5-30

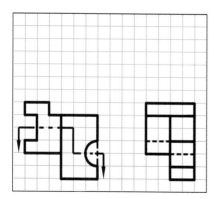

Figure P5-31

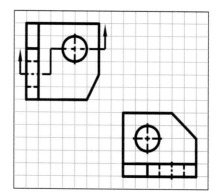

Figure P5-32

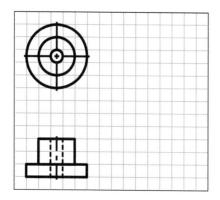

Figure P5-33

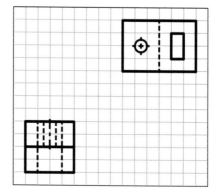

Figure P5-34

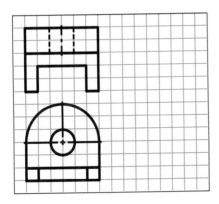

Figure P5-35

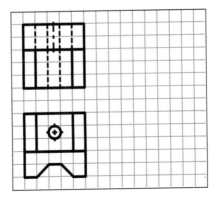

Figure P5-36

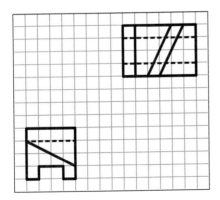

Figure P5-37

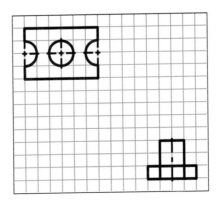

Figure P5-38

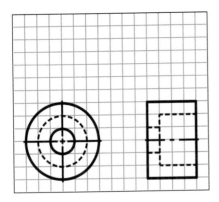

Figure P5-39

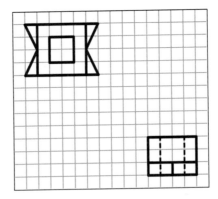

Figure P5-40

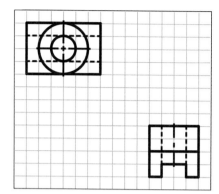

Figure P5-41

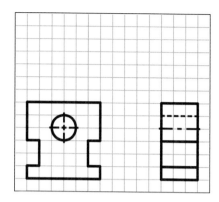

Figure P5-42

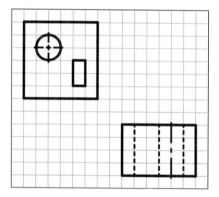

Figure P5-43

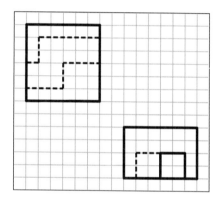

Figure P5-44

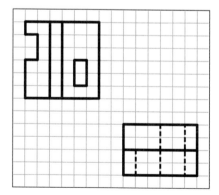

Figure P5-45

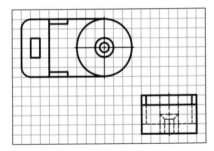

Figure P5-46

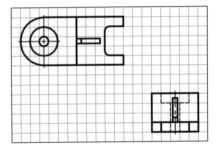

Figure P5-47

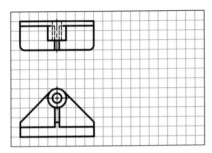

Figure P5-48

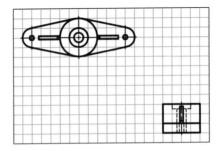

Figure P5-49

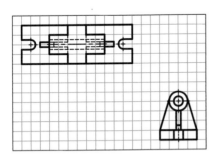

Figure P5-50

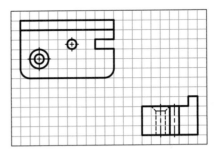

Figure P5-51

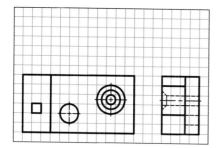

Figure P5-52

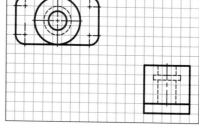

Figure P5-53

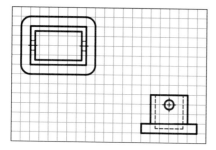

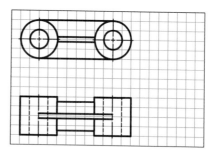

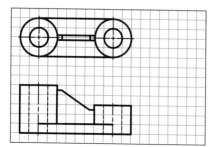

Figure P5-54

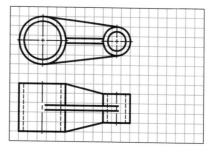

Figure P5-55

Figure P5-56

Figure P5-57

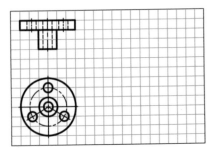

Figure P5-58

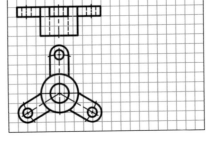

Figure P5-59

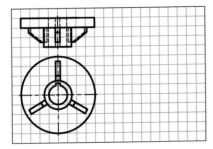

Figure P5-60

CHAPTER

6

DIMENSIONING AND TOLERANCING

☐ DIMENSIONING

Introduction

An engineering drawing, once submitted to production for manufacture or construction, must include all of the information needed to build the part, assembly, or system. To this end, technical drawings include dimensions and general notes describing the size and location of part features, as well as details related to the construction or manufacture of the part.

A *dimension* is a numerical value used to define the size, location, geometric characteristic, or surface texture of a part or feature. The main goals of dimensioning (as laid out in ANSI/ASME Y14.5M, Dimensioning and Tolerancing for Engineering Drawings), include the following:

1. Use only the dimensions needed to completely define the part, nothing more.

2. Select and arrange dimensions to support the function and mating relationship of the part. It is important that the dimensioned part not be subject to differing interpretations.

3. In general, do not specify the manufacturing methods to be used in building the part. This is done both to leave options open to manufacturing, and avoid potential legal problems.

4. Arrange the dimensions for optimum readability. Dimensions should appear in true profile views and refer to visible object edges.

5. Unless otherwise stated, assume angles to be 90 degrees.

Units of Measurement

Drawings are typically dimensioned using either millimeters or decimal inches. Metric (Système Internationale) drawings normally employ millimeters specified as whole numbers, as shown in Figure 6-1A. In the English or Imperial system, the preferred units are inches expressed in decimal form, as seen in Figure 6-1B. In some disciplines, notably architecture and construction, fractional inches are still employed, but decimals are preferred due to the easier arithmetic and greater precision that they provide. Drawings in inches are typically specified using two-decimal-place accuracy. In both metric and English drawings, there is no need to specify the units of individual dimensions. Rather, a general note similar to "Unless otherwise stated, all dimensions are in millimeters (or inches)" appears.

Application of Dimensions

Dimensions are applied to a drawing through the use of dimension lines, extension lines, and leaders from a feature to a dimension or note. In

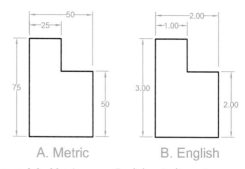

A. Metric   B. English

Figure 6-1   Metric versus English unit dimensions

addition, general notes are used to convey additional information.

TERMINOLOGY

Dimension lines are thin, dark lines used to show the direction and extents of a dimension. See Figure 6-2. A **dimension value** indicating the number of units of a measurement is associated with the dimension line. The height of the dimension value is typically 3 mm. By preference dimension lines are broken to allow the insertion of the dimension value. There is, however, an alternative dimension style that places the dimension value above an unbroken dimension line. Dimension lines are terminated by **arrows**, where the length of the arrowhead is equal to the dimension text height.

The dimension line of an angle being dimensioned is an arc drawn with its center at the apex of the angle.

As shown in Figure 6-2, thin, dark **extension lines** are typically drawn perpendicular to the associated dimension line. Extension lines extend from the view of an object feature to which they refer. A short visible gap (1.5 mm) is included between the extension line and the view for clarity. In addition, the extension line extends 3 mm beyond its outermost related dimension line.

Leader lines are drawn from a feature to a note, dimension, or symbol. As seen in Figure 6-2, leaders are inclined straight lines, except for a small horizontal shoulder that extends from to the mid-height of the first or last letter (or digit) of the note or dimension. These thin, dark lines start with an arrow at the feature being described. In situations where the feature is within an outline, a dot is placed and the arrow is drawn from it (see A36 steel leader line in Figure 6.2).

A **reference dimension** is only used for additional informational purposes. A reference dimension can be derived from other values shown on the drawing. It contains additional information, and is not used for production or inspection purposes. Reference dimensions are easily identified because the associated dimensional value is placed in parenthesis, as seen in Figure 6-2.

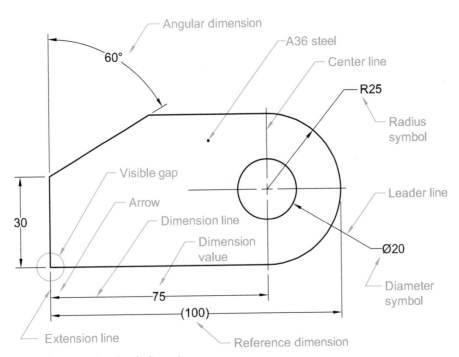

Figure 6-2 **Terminology associated with dimensions**

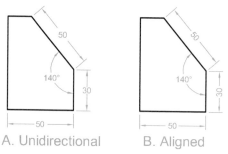

A. Unidirectional   B. Aligned

Figure 6-3 Unidirectional versus aligned dimensioning

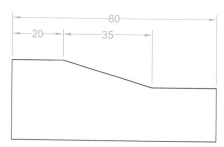

Figure 6-5 Grouping and alignment of parallel dimensions

Thin, dark *center lines* also play a role in dimensioning, since they are used to locate the centers of cylindrical parts and holes.

READING DIRECTION FOR DIMENSIONAL VALUES

In *unidirectional dimensioning*, dimension values and text are oriented horizontally, as shown in Figure 6-3A. In an older dimensioning style called *aligned dimensioning*, dimensional values are oriented parallel to their dimension lines, as shown in Figure 6-3B. Aligned dimensioning is not recognized by ANSI.

ARRANGEMENT, PLACEMENT, AND SPACING OF DIMENSIONS

As mentioned previously, dimensions are arranged for optimum readability. Several guidelines exist that govern the spacing, grouping, and staggering of parallel dimensions. There are also guidelines for dimensioning when space is limited.

A distance of at least 10 mm between the first dimension line and the part should be maintained. For succeeding parallel dimensions, this distance should be at least 6 mm (see Figure 6-4).

Parallel (i.e., either horizontal, vertical, or aligned) dimensions should be grouped and aligned, as shown in Figure 6-5, in order to present a uniform appearance.

The dimensional values of parallel dimensions should be staggered, as shown in Figure 6-6, to avoid crowding.

By preference the dimensional value and arrows should appear inside the extension lines. Depending upon the available space, however, it may be necessary to leave only the dimensional value inside, only the arrows inside, or nothing inside. See Figure 6-7 on page 162 for these possibilities. Note that this situation applies to horizontal, vertical, aligned, angular, and radial dimensions.

There are also guidelines for leaders that aim to improve the readability of a drawing. For instance, multiple leaders that are close to one another should be drawn parallel, as shown in Figure 6-8 on page 162. Leader lines should not be overly long, and they cross as few lines as possible. Leader lines should never cross one another. Finally, leaders directed to a circle or arc, as

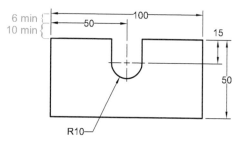

Figure 6-4 Spacing between parallel dimensions

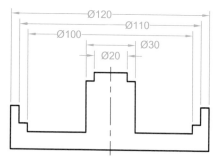

Figure 6-6 Staggering of parallel dimensions

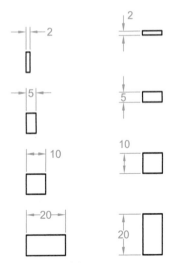

Figure 6-7 Placement of dimension text

shown in Figure 6-9, should be radial (i.e., pass through the center, if extended) to the hole or arc.

Using Dimensions to Specify Size and Locate Features

Dimensions are used to specify the size and location of features. Features are sized and located with linear (horizontal, vertical, aligned), radial, diametric, and angular dimensions. Figure 6-10

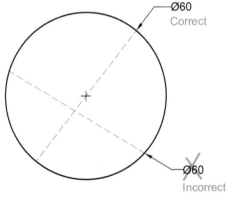

Figure 6-9 Leaders directed to circle or arc should pass through circle center when extended

provides a drawing with dimensions used to size the part features, while Figure 6-11 shows the same drawing, but with the dimensions used to locate part features now displayed.

Symbols, Abbreviations, and General Notes

A number of symbols are employed in association with dimensioning. Figure 6-12 shows several of these symbols, including radius, diameter, spherical radius, spherical diameter, counterbore, countersink, deep, and times.

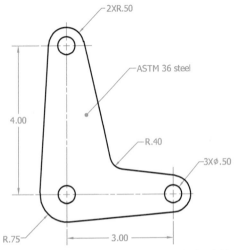

Figure 6-8 Multiple leaders in same vicinity should be parallel

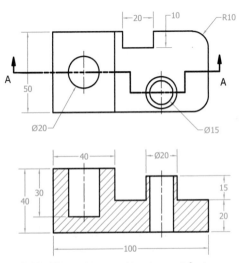

Figure 6-10 Dimensions used to size part features

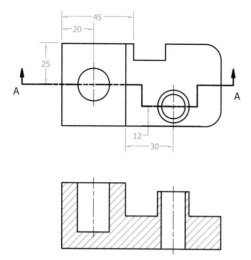

Figure 6-11   Dimensions used to locate part features

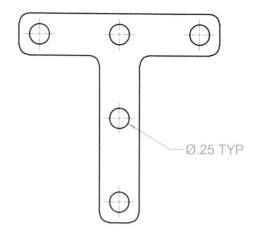

NOTE: All fillets and rounds R.125

Figure 6-13   General notes and abbreviations used in dimensioning

Whenever several features (e.g., holes, fillets, rounds) of the same type and size appear in a drawing, either a general note or the abbreviation TYP (for typical) may be used, as shown in Figure 6-13. Also note that the *times* symbol (i.e., $2 \times R10$) is also used to dimension multiple features of the same size (for example, Figure 6-8 on page 162).

Dimensioning Rules and Guidelines

In this section several dimensioning rules or guidelines will be discussed. Note that on occasion these rules may be violated because of part complexity, lack of space, conflict with other rules, etc. Rules concerning prismatic shapes will first be covered, followed by rules concerning cylinders and arcs.

PRISMS

1. *Do not repeat dimensions.* The depth of the object shown in Figure 6-14 is 30. Although this dimension appears in the top view, it could just as well have been placed on the right side view. In no case, however, should this same dimension appear in both views.

2. *Apply dimensions to a feature in its most descriptive view.* The object shown in Figure 6-14 contains a single (extruded "L" shape)

Symbol Name	Symbol
Counterbore	⌴
Countersink	∨
Deep	⊽
Diameter	Ø
Square	□
Places, Times	X
Radius	R
Spherical Radius	SR
Spherical Diameter	SØ

Figure 6-12   Dimensioning symbols

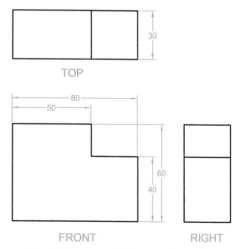

Figure 6-14   Rules and guidelines for dimensioning prisms

feature. The most descriptive view of this feature is the front view. Notice that four of the five dimensions needed to fully constrain the object appear on the front view.

3. *Dimension between views.* The object shown in Figure 6-14 contains two width dimensions (50 and 80) and two height dimensions (40 and 60). Since the width is shared by the front and top views, these width dimensions are placed between these views. Likewise the height dimensions are placed between the front and right views.

4. *Omit one (intermediate) dimension.* The object in Figure 6-14 contains two intermediate width dimensions, 50 and 30 (since 80−50 = 30). Only one of them is shown. Likewise, there are two intermediate height dimensions, 40 and 20 (since 60 − 40 = 20). Only one of them is shown. This is done to avoid cluttering the drawing with unnecessary dimensions, and also to avoid ambiguity in specifying tolerances (see the section on tolerance accumulation later in the chapter).

5. *Place smaller dimensions inside larger dimensions.* Note that in Figure 6-14 the intermediate dimensions (50, 40) are placed closer to the view than the principal dimensions (80, 60). This practice helps to keep the drawing organized by avoiding the need for extension lines that cross dimension lines.

Two additional features, an extruded cut and a hole, have been added to the object depicted in Figure 6-14, and can be seen in Figure 6-15.

6. *Dimension to visible object lines, not to hidden lines.* The dimensions of the newly added cut, 30 and 10, are placed on the view that best describes this feature (i.e., the top view), and not in either the front (30) or the right (10) views, where it would have been necessary to apply a dimension to a hidden line.

7. *Keep dimensions outside the views.* The diameter of the through hole should be placed outside the view. If the drawing view is cluttered or the leader line needs to be extremely

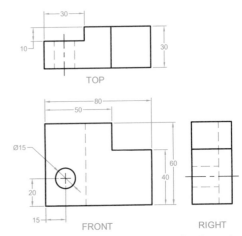

Figure 6-15 More rules and guidelines for dimensioning prisms

long to place the dimension outside the view, then this rule may be overridden.

8. *Extension lines may cross object lines and other extension lines.* In general it is desirable to avoid lines that cross; however, it is permissible for an extension line to cross an object line (e.g., extension lines for the dimensions of 20 and 15 locating the hole center), or another extension line (80 and 60).

CYLINDERS AND ARCS

As shown in Figure 6-16, calipers are used to directly measure the diameter of solid cylinders

Figure 6-16 Use of calipers to measure the diameter of a cylinder

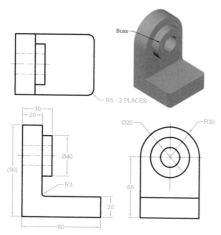

Figure 6-17 Rules and guidelines for dimensioning cylinders and arcs

and round holes. It is for this practical reason that the diameter rather than the radius of circular features is specified on engineering drawings.

9. *Dimension the diameter of cylindrical parts in their rectangular view.* In Figure 6-17 the diameter of the boss (raised cylindrical) feature is dimensioned in its rectangular (i.e., front) view. Note that the diameter symbol ø precedes the dimension.

10. *Dimension the diameter of cylindrical holes in their circular view.* In Figure 6-17 the diameter of the through hole is dimensioned in the right view, where the hole appears as a circle.

11. *Dimension the radius of circular arcs in the view where their true shape is seen.* Note that the symbol R precedes the dimensional value of the radius. As seen in Figure 6-17, this rule applies to fillets (R3), rounds (R5), and other circular arcs (R30). For arcs less than or equal to 180 degrees, specify the radius. For arcs greater than 180 degrees, specify the diameter.

Some additional comments regarding Figure 6-17 are in order. First, note that the use of the note PLACES (i.e., R5–2 PLACES) to eliminate the need for an additional R5 dimension. Sim-

ilarly, the abbreviation TYP is also common. Also notice that the R30 radial dimension eliminates the need to dimension the overall depth (60) of the object. Similarly, the overall height (90) is not needed, although it is provided in the form of a reference dimension.

Finally, notice that the overall width dimension (80) appears at the bottom of the front view, in apparent violation of Rule 3, by which the width dimensions would appear between the front and top views. This is due to yet another guideline:

12. *Avoid overly long extension and leader lines.* By placing the 80 width dimension at the bottom, rather than the top, of the front view, the length of the associated extension line is significantly reduced.

□ TOLERANCING

Introduction

In manufacturing, the same process is typically employed to mass produce a single part. These parts are then combined with other similarly mass-produced parts to create commercial products. Clearly these mass-produced parts must be interchangeable. However, when inspecting a batch of parts produced by the same manufacturing process, we would not find two parts that are exactly the same. In even the most precise manufacturing process, slight variations in part size are found.

Tolerancing is a dimensioning technique used to ensure part interchangeability by controlling the variance in manufactured parts. This is accomplished by specifying a range within which a dimension is allowed to vary. As long as the size and location of part features fall within this tolerance zone, the part should function properly within an assembly.

Tolerancing is critical to the success of manufacturing. Beyond ensuring the interchangeability of parts, tolerancing directly influences both the cost and quality of manufactured parts. Parts that are made to high accuracy are expensive.

Depending upon the type of product, extremely accurate parts may not be warranted. For example, the parts used to make a plastic toy do not need to be as accurate as automotive parts. As a general rule, tolerances should be stated as generously as possible, while still allowing the part to function properly. Doing so allows for the possibility of using a wider variety of processes to manufacture the part, and consequently helps keep part costs low.

Manufacturing quality is primarily a function of part accuracy. High-quality parts exhibit small variations in size and shape. By specifying tight tolerance zones, and then controlling part variability using techniques like **statistical process control**, the designer can maintain and even improve product quality.

Definitions

A **tolerance** is the total permissible variation of a size, or the difference between the maximum and minimum **limits of size**. The tolerance 3.25 ± 0.03 indicates that the actual part size can range anywhere between 3.22 and 3.28 and still function properly. In this example, the value 3.28 is the maximum limit of size, 3.22 is the minimum limit of size, and the tolerance is 0.06.

While the **actual size** is the measured size of a finished part, the **basic size** is the theoretical size from which a tolerance is assigned. In the example cited above, 3.25 is the basic size, and the actual size will fall between 3.22 and 3.28, assuming that it is within tolerance.

Tolerance Declaration

Tolerances may be expressed in different ways, including:

1. Direct tolerancing methods
2. General tolerance notes
3. Geometric tolerances

Direct tolerancing methods include 1) limit dimensioning and 2) plus and minus tolerancing. In limit dimensioning, the limits of size are directly represented as part of the dimension. The upper limit (maximum value) is placed above the lower

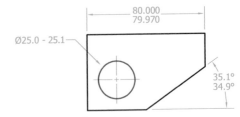

Figure 6-18 Limit dimensioning

limit (minimum value). When expressed in a single line the lower limit precedes the upper limit. Figure 6-18 provides some limit dimensioning examples.

In plus and minus tolerancing, the basic dimension is given first, followed by a plus and minus expression of the tolerance. Plus and minus tolerances may either be **unilateral** or **bilateral**. A unilateral tolerance is only permitted to vary in one direction from the basic size. Bilateral tolerances, on the other hand, may vary in either direction from the basic size. The variance of bilateral dimensions may either be equal or unequal. Figure 6-19 provides some plus and minus tolerance dimensioning examples.

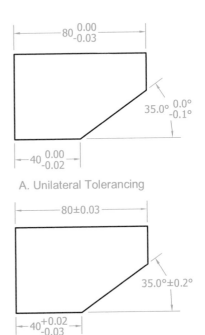

A. Unilateral Tolerancing

B. Bilateral Tolerancing

Figure 6-19 Plus and minus dimensioning

General notes like "ALL DIMENSIONS HELD TO ±0.05" are sometimes used on engineering drawings. Such a note indicates that all dimensions appearing on the drawing should fall within 0.05 inches of the basic size.

Figure 6-20 provides an example of an object that has been dimensioned using geometric dimensioning and tolerancing (GD&T) techniques. GD&T is covered in Appendix B of this text.

Tolerance Accumulation

When the location of a feature depends upon more than one tolerance value, these tolerances will be cumulative. In the **chain dimensioning** technique employed in Figure 6-21A on page 168, for example, the tolerance accumulation between surfaces X and Y is ±0.03. The tendency for tolerances to stack up or accumulate can be reduced by using **base line dimensioning**, as shown in Figure 6-21B. In this case, the tolerance variation between surfaces X and Y is reduced to ±0.02. If necessary, the tendency for tolerances to

accumulate can be further controlled by **direct dimensioning**. The maximum variation between the directly dimensioned surfaces X and Y in Figure 6-21C is ±0.01.

Mated Parts

The tolerance of a single standalone part is of little importance. When a part is mated with other parts in an assembly the true value of tolerancing becomes apparent. Mated parts must be toleranced as a system to fit within a prescribed degree of accuracy. Figure 6-22 on page 168, for example, shows a shaft, bushing, and pulley assembly. The shaft must be able to turn freely within the bushing, while the bushing is force-fit into the pulley.

Fit refers to the degree of tightness or looseness between two mating parts. In a **clearance fit**, the internal member (e.g., shaft) is always smaller than the external member (e.g., hole). The shaft and bushing in Figure 6-22 above have a clearance fit. The shaft is free to turn inside the hole.

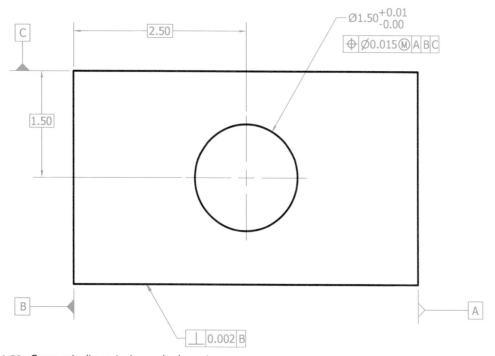

Figure 6-20 **Geometric dimensioning and tolerancing**

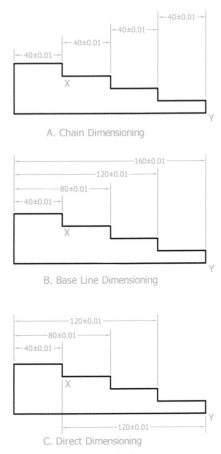

A. Chain Dimensioning

B. Base Line Dimensioning

C. Direct Dimensioning

Figure 6-21 Accumulation of tolerances

In an *interference fit*, the internal member is always larger than the external member. An interference fit requires that the two parts be forced together, as is the case with the bushing and pulley shown in Figure 6-22. Note that an interference fit fastens two parts together without using adhesive or mechanical fasteners.

A *transition fit* ranges between a pure clearance and a pure interference fit. In a transition fit either the internal shaft or the external hole may be larger, so that parts either slide or are forced together. In a *line fit*, one of the limits on both the hole and the shaft are equal, meaning that the shaft and hole may have the same size.

Figure 6-23 shows examples of clearance, interference, transition, and line fits, along with their upper and lower limits. Later in this chapter

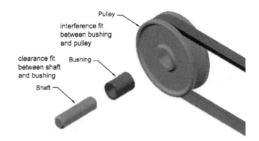

Figure 6-22 Pulley assembly tolerance fits

we will see that each of these classes of fit can be further categorized into subclasses.

The *allowance* is the tightest possible fit between two mated parts. It is the difference between the smallest hole size and the largest shaft size. For a clearance fit, the allowance is positive and represents the minimum clearance between the two parts. For an interference fit, the allowance is negative and represents the maximum interference between the two parts. The allowance is calculated for the four cases shown in Figure 6-23.

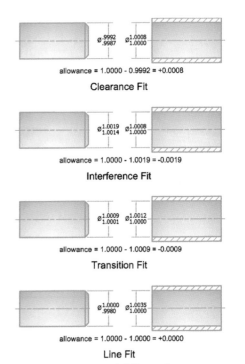

Figure 6-23 Comparison of different fit types

While these different types of fit typically refer to cylindrical features like shafts and holes, they also apply to parts with parallel surfaces that fit inside one another, as depicted in Figure 6-24.

With an understanding of tolerance, allowance, basic size, and types of fit, we can assign tolerances to a system of mated parts in order to achieve a particular type of fit (i.e., clearance, interference). All that is needed is a reference system, or method of calculation, for relating the tolerances and allowance to the basic size. In the following sections, two reference systems, hole and shaft, are discussed. Both English and metric units, while calculated differently, employ hole and shaft methods of calculation.

Basic Hole System: English Units

In the basic hole system the minimum (i.e., lower limit) hole size is taken as the basic size. The allowance between the hole and shaft is then determined, and the tolerances are applied. The basic hole system is widely used because of the ready availability of tools (e.g., drills, reamers) capable of producing standard size holes with precision. In effect, when using the basic hole system, we are choosing a standard drill size to create the hole, and then turning down the shaft to fit that hole. Figure

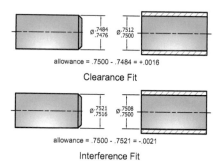

allowance = .7500 - .7484 = +.0016

Clearance Fit

allowance = .7500 - .7521 = -.0021

Interference Fit

Figure 6-25 Basic hole system for clearance and interference fits; basic size = .7500

6-25 illustrates the basic hole system for both a clearance and an interference fit.

Step-by-step tolerance calculation of a clearance fit using the basic hole system

Given:

- Basic size is .5000
- Allowance is +.0020
- Hole tolerance is .0016
- Shaft tolerance is .0010

1. Hole minimum = basic size = .5000

2. To find the upper limit on the shaft:
   Since the
   Allowance = Hole minimum − Shaft maximum,
   Shaft maximum = Hole minimum − Allowance = .5000 − .0020 = .4980

3. To find the upper limit on the hole:
   Hole maximum = Hole minimum + Hole tolerance = .5000 + .0016 = .5016

4. To find the lower limit on the shaft:
   Shaft minimum = Shaft maximum − Shaft tolerance = .4980 − .0010 = .4970

   See Figure 6-26.

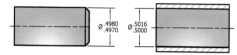

Figure 6-26 Step-by-step tolerance calculation: clearance, basic hole

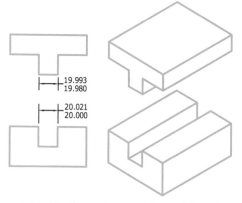

Figure 6-24 Fit of mated parts with parallel surfaces

Step-by-step tolerance calculation of an interference fit using the basic hole system

Given:

- Basic size is 4.0000
- Allowance is −.0049
- Hole tolerance is .0014
- Shaft tolerance is .0009

1. Hole minimum = basic size = 4.0000

2. To find the upper limit on the shaft:
   Shaft maximum = Hole minimum−Allowance = 4.0000−(−.0049) = 4.0049
   (Since Allowance = Hole minimum−Shaft maximum)

3. To find the upper limit on the hole:
   Hole maximum = Hole minimum + Hole tolerance = 4.0000 + .0014 = 4.0014

4. To find the lower limit on the shaft:
   Shaft minimum = Shaft maximum−Shaft tolerance = 4.0049−.0009 = 4.0040

See Figure 6-27.

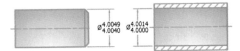

Figure 6-27 Step-by-step tolerance calculation: interference, basic hole

Basic Shaft System: English Units

Less common than the basic hole system is the basic shaft system. In the basic shaft system the maximum (i.e., upper limit) shaft is taken as

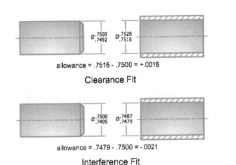

allowance = .7516 - .7500 = +.0016

Clearance Fit

allowance = .7479 - .7500 = -.0021

Interference Fit

Figure 6-28 Basic shaft system for clearance and interference fits; basic size = .7500

the basic size. When using the basic shaft system, a stock shaft is selected, and the mating hole is created to suit the shaft. The basic shaft system should only be used when there is a specific reason for using it, as for example when a shaft cannot be easily machined to size, or when several parts requiring different fits must be mated to the same shaft. Figure 6-28 illustrates the basic shaft system for clearance and interference fits.

Step-by-step tolerance calculation of a clearance fit using the basic shaft system

Given:

- Basic size is 8.0000
- Allowance is +.0150
- Hole tolerance is .0120
- Shaft tolerance is .0070

1. Shaft maximum = basic size = 8.0000

2. To find the lower limit on the hole:
   Hole minimum = Shaft maximum + Allowance = 8.0000 + .0150 = 8.0150
   (Since Allowance = Hole minimum−Shaft maximum)

3. To find the upper limit on the hole:
   Hole maximum = Hole minimum + Hole tolerance = 8.0150 + .0120 = 8.0270

4. To find the lower limit on the shaft:
   Shaft minimum = Shaft maximum−Shaft tolerance = 8.0000−.0070 = 7.9930

See Figure 6-29.

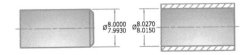

Figure 6-29 Step-by-step tolerance calculation: clearance, basic shaft

Preferred English Limits and Fits

In order to simplify the process of tolerancing mated parts, ANSI standards and accompanying tables have been developed for both English and metric units. The English unit standards are described in B4.1–1967 (R1994), "Preferred Limits and Fits for Cylindrical Parts." Although these standards are intended for holes, cylinders, and shafts, they can also be used for fits between parallel surfaces (see Figure 6-24 on page 169). The tables taken from B4.1 appear in Appendix C. To use these tables, the user provides the basic size and the type of fit. These tables employ the basic hole system.

B4.1 recognizes five different types of fit, as well as different classes within each type of fit. For any one class of fit (e.g., RC5), the fit produced between the mated parts results in the same fit characteristics, regardless of the basic size of the part features. The characteristics of the different types and classes of fit are described below.

RUNNING OR SLIDING CLEARANCE FIT (RC)

Clearance fits that provide a similar running performance throughout a range of sizes. The clearances of the first two classes (RC1, RC2) are intended for use as slide fits, while the other classes (RC3 through RC9) are for free-running operation. RC1 and RC2 clearances increase more slowly with diameter than the other classes to maintain accurate location at the expense of free relative motion. The other (free-running) classes range from precision (RC3) to loose (RC9).

LOCATIONAL CLEARANCE FIT (LC)

Tighter clearance fits than RC, intended for parts that are normally stationary but can be freely assembled and disassembled. They range from snug-line fits for parts requiring accuracy of location to looser fastener fits where freedom of assembly is of prime importance.

TRANSITION CLEARANCE OR INTERFERENCE FIT (LT)

Compromise between clearance and interference fits, for application where accuracy of location is important, but a small amount of either clearance or interference is permissible.

LOCATIONAL INTERFERENCE FIT (LN)

Force fit used where accuracy of location is of prime importance and for parts requiring rigidity

and alignment with no special requirements for bore pressure. Not intended for parts designed to transmit frictional loads from one part to another based on tightness of fit.

FORCE OR SHRINK FIT (FN)

Force fit characterized by maintenance of constant bore pressures throughout the range of sizes. The interference varies almost directly with diameter, with the difference between its maximum and minimum values kept small to maintain the resulting pressures within reasonable limits.

> **Step-by-step tolerance calculation using English unit fit tables, basic hole system**

Given:

- Basic size = 2.0000
- Fit type is RC8
- Calculation method is basic hole

1. From Appendix C, Running and Sliding Fits, with a basic size of 2.0000 and an RC8 fit type, the following information is extracted:

Nominal Size Range (Inches)	Class RC8		
	Limits of Clearance	Standard Limits	
Over To		Hole H10	Shaft c9
1.97 − 3.15	6	+4.5	–6.0
	13.5	0	–9.0

A note at the top of the table indicates that the limits are in thousandths of an inch. Thus the upper and lower limits on the hole, +4.5 and 0, are actually .0045 and 0, while the limits on the shaft, −6.0 and −9.0, are −.0060 and −.0090. These standard limits are added algebraically to the basic size to determine the actual tolerance limits. The limits of clearance give the tightest (0−(−.0060) = +.0060) and the loosest $(+.0045-(-.0090) = +.0135)$ possible fits. Recall that the tightest fit is the allowance.

2. Tolerance limits on the hole:
   Upper limit $= 2.0000 + .0045 = 2.0045$
   Lower limit $= 2.0000$ ($=$ basic size)

3. Tolerance limits on the shaft:
   Upper limit $= 2.0000-.0060 = 1.9940$
   Lower limit $= 2.0000-.0090 = 1.9910$

4. Allowance $=$ Hole minimum$-$Shaft maximum $=$ $2.0000-1.9940 = .0060$

   See Figure 6-30.

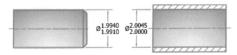

Figure 6-30 Step-by-step tolerance calculation: LC3, basic hole, preferred English fit tables

Step-by-step tolerance calculation using English units fit tables, basic shaft

Given:

- Basic size is 1.0000
- Fit type is LN2
- Calculation method is basic shaft

ANSI Standard B4.1 only provides preferred limits and fits tables using the Basic Hole System. To calculate tolerances of mated parts using the Basic Shaft System, the standard limits must be converted from basic hole to basic shaft.

1. From Appendix C, Locational Interference Fits, with a basic size of 1.0000 and an LN2 fit type, the following information is extracted:

Nominal Size Range (Inches)	Class LN2		
	Limits of Clearance	Standard Limits	
Over To		Hole H7	Shaft p6
0.71 − 1.19	0	+0.8	+1.3
	1.3	–0	+0.8

2. These standard limits are using the basic hole system. In the basic shaft system, the upper limit on the shaft is taken as the basic size, meaning that the upper limit on the shaft should be 0, not +1.3. We can therefore convert the standard limits from basic hole to basic shaft by subtracting + 1.3 from all of the standard limits:

	Hole	Shaft
Upper limit	$+0.8 - 1.3 = -0.5$	$+1.3 - 1.3 = 0$
Lower limit	$0 - 1.3 = -1.3$	$+0.8 - 1.3 = -0.5$

3. Tolerance limits on the hole:
   Upper limit $= 1.0000-.0005 = .9995$
   Lower limit $= 1.0000-.0013 = .9987$

4. Tolerance limits on the shaft:
   Upper limit $= 1.0000$ ($=$ basic size)
   Lower limit $= 1.0000-.0005 = .9995$

5. Allowance $=$ Hole minimum$-$Shaft maximum $=$ $.9987-1.0000 = -.0013$

   See Figure 6-31.

Figure 6-31 Step-by-step tolerance calculation: LN2, basic shaft, preferred English fit tables

Preferred Metric Limits and Fits

ANSI B4.2–1978 (1994), "Preferred Metric Limits and Fits," provides standards and tables for tolerancing fitted parts using metric units. Using this standard, a tolerance is specified using a special designation, for example 40H7. ANSI B4.2 begins with a series of definitions and an accompanying illustration similar to the one shown in Figure 6-32 on page 173. These definitions include:

Basic size—The size to which limits or deviations are assigned. It is designated by the number 40 in 40H7.

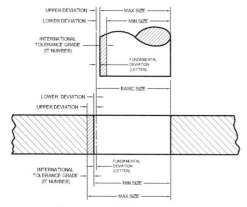

Figure 6-32   Illustration of definitions for metric limits and fits

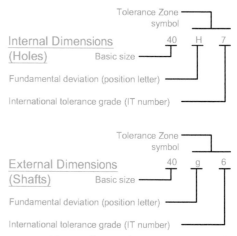

Figure 6-33   Metric-toleranced size with associated terminology

Deviation—The algebraic difference between a size and the corresponding basic size.

Upper deviation—The algebraic difference between the maximum limit of size and the corresponding basic size.

Lower deviation—The algebraic difference between the minimum limit of size and the corresponding basic size.

Fundamental deviation—The deviation, upper or lower, that is closest to the basic size. It is designated by the letter H in 40H7.

Tolerance—The difference between the maximum and minimum size limits on a part.

Tolerance zone—A zone representing the tolerance and its position in relation to the basic size.

International tolerance grade (IT)—A group of tolerances that vary depending on the basic size but provide the same relative accuracy within a given grade. It is designated by the number 7 in 40H7 (IT7).

Hole basis—The system of fits where the minimum hole size is equal to the basic size. The fundamental deviation for a hole-basis system is "H."

Shaft basis—The system of fits where the maximum shaft size is equal to the basic size. The fundamental deviation for a shaft-basis system is "h."

Figure 6-33 shows a toleranced size, along with the associated terminology. The International Tolerance grade establishes the magnitude of the tolerance zone (i.e., the amount of part size variation that is allowed) for both internal (hole) and external (shaft) dimensions. It is expressed in grade numbers (e.g., IT7), with smaller grade numbers indicating a smaller tolerance zone.

The Fundamental deviation establishes the position of the tolerance zone with respect to the basic size. It is expressed by "tolerance position letters," with upper-case letters (e.g., H) being used for hole dimensions and lower-case letters (e.g., h) used for shaft dimensions.

A tolerance symbol (e.g., H7) is formed by combining IT grade number and the tolerance position letter. The tolerance symbol identifies the actual maximum and minimum limits of the part. Toleranced sizes (e.g., 40H7) are determined by the basic size followed by a tolerance symbol.

A fit (e.g., 40 H8/f7) between mated parts is indicated by the basic size common to both components, followed by a symbol corresponding to each component, with the internal (hole) part symbol preceding the external (shaft) part symbol. Figure 6-34 on page 174 shows a fit as designated using B4.2.

Fit Designation   40   H7 / g6

Basic size ─────┘

Fit ──────┘

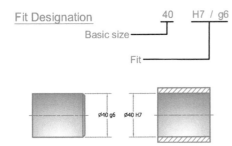

Figure 6-34 Metric-fit designation

Standard, or preferred, sizes of round metal parts should be used whenever possible. Table 6-1 shows these preferred basic sizes. The basic size of mating parts should, when possible, be chosen from the first-choice sizes listed in this table.

As with English fits, metric preferred fits are based either on hole or on shaft parts. Preferred fits to relative scale are shown in Figure 6-35 for hole-basis and Figure 6-36 on page 176 for shaft-basis fits. Hole basis fits have a fundamental deviation of "H" on the hole, while shaft basis fits have a fundamental deviation of

"h" on the shaft. While in most cases hole basis is the preferred system, shaft basis should be used when a common shaft mates with different holes.

Figure 6-37 on page 177 provides a description of hole and shaft basis fits that have the same relative fit condition. In the table appearing in Appendix D, metric limits and fits of clearance, transition, and interference fits are provided. Appendix D uses the fit types described in Figure 6-37 and the preferred sizes appearing in Table 6-1.

Step-by-step tolerance calculation using metric fit tables, hole basis

Given:

- Basic size is 50
- Fit type is free running, H9/d9
- Calculation method is hole basis

1. Entering the table in Appendix D, Preferred Hole Basis Clearance Fits, with a basic size of 50 and a free-running H9/d9 fit type, the following information is extracted:

BASIC SIZE		FREE RUNNING		
		Hole H9	Shaft d9	Fit
50	MAX	50.062	49.920	0.204
	MIN	50.000	49.858	0.080

2. Tolerance limits on the hole:
   Upper limit = 50.062
   Lower limit = 50.000 (= basic size)

3. Tolerance limits on the shaft:
   Upper limit = 49.920
   Lower limit = 49.858

4. Allowance = Hole min−Shaft max =
   50.000−49.920 = +0.080

   See Figure 6-38 on page 177.

Table 6-1

Basic Size, mm		Basic Size, mm		Basic Size, mm	
First Choice	Second Choice	First Choice	Second Choice	First Choice	Second Choice
1		10		100	
	1.1		11		110
1.2		12		120	
	1.4		14		140
1.6		16		160	
	1.8		18		180
2		20		200	
	2.2		22		220
2.5		25		250	
	2.8		28		280
3		30		300	
	3.5		35		350
4		40		400	
	4.5		45		450
5		50		500	
	5.5		55		550
6		60		600	
	7		70		700
8		80		800	
	9		90		900
				1000	

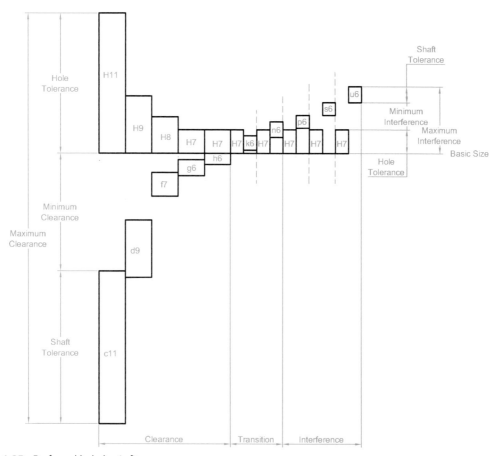

Figure 6-35   Preferred hole basis fits

Given:

- Basic size is 30
- Fit type is medium drive, S7/h6
- Calculation method is shaft basis

1. Entering the table in Appendix D, Preferred Shaft Basis Interference Fits, with a basic size of 30 and a medium drive S7/h6 fit type, the following information is extracted:

BASIC SIZE		MEDIUM DRIVE		
		Hole S7	Shaft h6	Fit
30	MAX	29.973	30.000	−0.014
	MIN	29.952	29.987	−0.048

2. Tolerance limits on the hole:
   Upper limit = 29.973
   Lower limit = 29.952

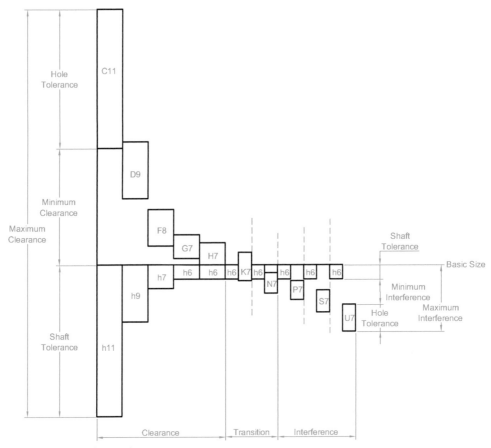

Figure 6-36 Preferred shaft basis fits

3. Tolerance limits on the shaft:
   Upper limit = 30.000 (= basic size)
   Lower limit = 29.987

4. Allowance = Hole min−Shaft max =
   29.952−30.000 = −0.048

   See Figure 6-39 on page 177.

Tolerancing in CAD

CAD programs typically provide a variety of tools for the specification of toleranced dimensions within a drawing. The dialog box shown in Figure 6-40 on page 177, for example, allows the designer to display tolerances in several ways. The dialog box settings shown in Figure 6-40 on the left result in the toleranced dimension shown on the right.

Finish Marks

Parts formed by casting have rough external surfaces. When these cast parts are used in an assembly, surfaces in contact with other parts are machined or finished, in order to provide a smooth mating surface, reduce friction, etc. In an engineering drawing a finish mark symbol ($\sqrt{}$) is used to indicate that a surface is to be machined. As seen in Figure 6-41 on page 178, finish marks are applied to all edge views, whether visible or hidden, of finished part surfaces.

ISO SYMBOL		DESCRIPTION
Hole Basis	**Shaft Basis**	
H11/c11	C11/h11	Loose running fit for wide commercial tolerances or allowances on external members.
H9/d9	D9/h9	Free running fit not for use where accuracy is essential, but good for large temperature variations, high running speeds, or heavy journal pressures.
H8/f7	F8/h7	Close running fit for running on accurate machines and for accurate location at moderate speeds and journal pressures.
H7/g6	G7/h6	Sliding fit not intended to run freely, but to move and turn freely and locate accurately.
H7/h6	H7/h6	Locational clearance fit provides snug fit for locating stationary parts, but can be freely assembled and disassembled.
H7/k6	K7/h6	Locational transition fit for accurate location, a compromise between clearance and interference.
H7/n6	N7/h6	Locational transition fit for more accurate location where grater interference is permissible.
H7/p6	P7/h6	Locational interference fit for parts requiring rigidity and alignment with prime accuracy of location but without special bore pressure requirements.
H7/s6	S7/h6	Medium drive fit for ordinary steel parts or shrink fits on light sections, the tightest fit usable with cast iron.
H7/u6	U7/h6	Force fit suitable for parts which can be highly stressed or for shrink fits where the heavy pressing forces required are impractical.

Clearance Fits / Transition Fits / Interference Fits — More Clearance / More Interference

Figure 6-37   Description of preferred fits

Figure 6-38   Step-by-step tolerance calculation: free running, hole basis, preferred metric fit tables

Figure 6-39   Step-by-step tolerance calculation: medium drive, shaft basis, preferred metric fit tables

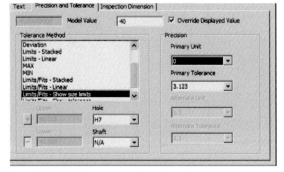

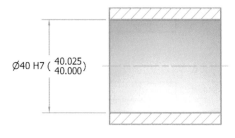

Figure 6-40   Tolerance specification using CAD (Courtesy of Autodesk, Inc.)

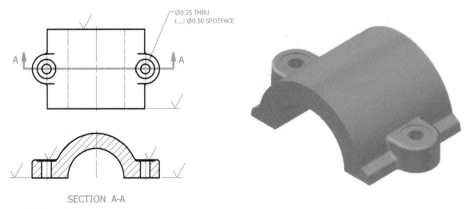

Ø0.25 THRU
⌴ Ø0.50 SPOTFACE

SECTION A-A

Figure 6-41   Finish marks

☐ QUESTIONS

DIMENSIONING

1. Construct all extension lines, dimension lines, leaders, and arrowheads that are necessary to fully dimension the objects. It is not necessary to include the actual dimension values (P6-1 through P6-8).

2. Select the dimensions A-Z that represent good dimensioning (P6-9 through P6-10). When finished, the object should be fully dimensioned. In cases where two dimensions locate or size the same feature, accept the one that best meets good dimensioning practice guidelines, while rejecting the other.

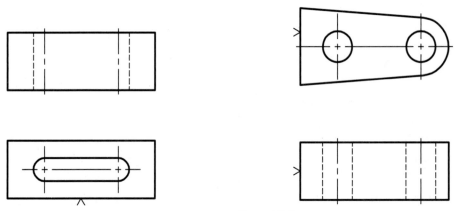

Figure P6-1   Figure P6-2

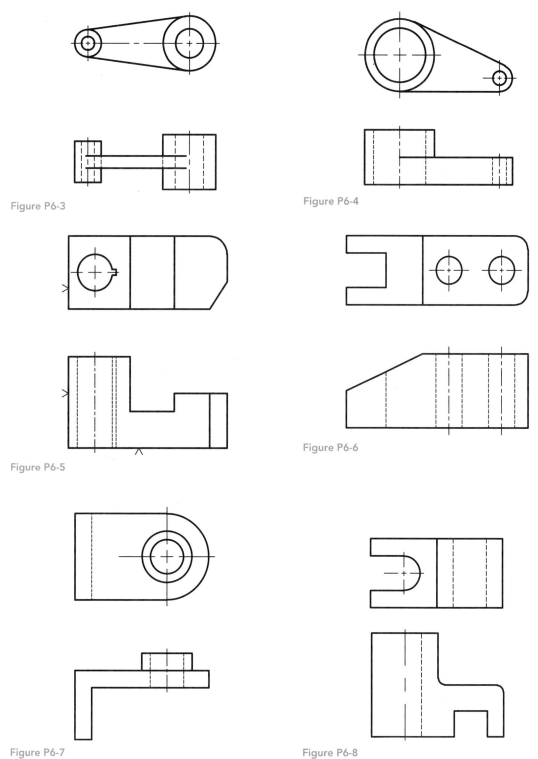

Figure P6-3

Figure P6-4

Figure P6-5

Figure P6-6

Figure P6-7

Figure P6-8

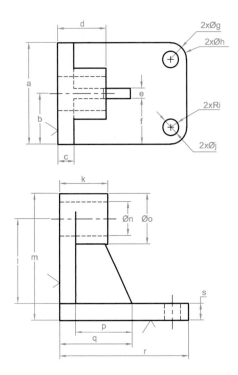

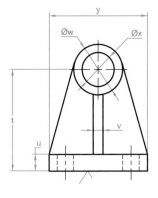

Figure P6-9

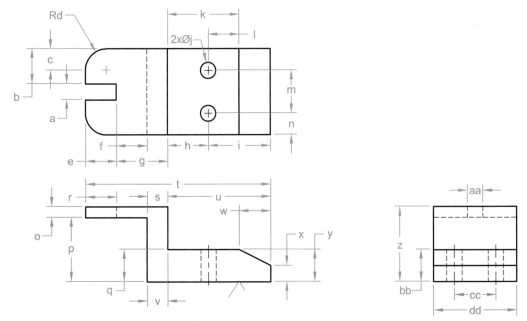

Figure P6-10

TOLERANCING

1. Given the method of calculation, basic size, fit (i.e., clearance, interference, or transition), tolerances, and the allowance, determine the limits on the hole and shaft.

	a.	b.	c.	d.	e.
Basic Size	10.000	1.5000	0.5000	30.000	16.000
Class of Fit	Clearance	Interference	Clearance	Interference	Transition
Method of Calculation	Shaft	Hole	Hole	Shaft	Hole
Limits of Hole Size					
Limits of Shaft Size					
Tolerance on Hole	0.200	0.0015	0.0010	0.016	0.025
Tolerance on Shaft	0.100	0.0014	0.0007	0.014	0.024
Allowance	0.150	–0.0030	0.0017	–0.032	–0.016

2. Using the appropriate limit dimensioning tables and either the basic hole or basic shaft method of calculation, complete the portions of the table.

	a.	b.	c.	d.
Basic Size	2.5000	1.5000	4.2500	10.0000
Class of Fit	RC1	LT4	LN1	FN2
Method of Calculation	Basic Hole	Basic Shaft	Basic Hole	Basic Hole
Limits of Hole Size				
Limits of Shaft Size				
Tolerance on Hole				
Tolerance on Shaft				
Allowance				

	e.	f.	g.	h.
Basic Size				3.7500
Class of Fit	RC7	LC8	LN3	FN5
Method of Calculation	Basic Hole	Basic Hole	Basic Shaft	
Limits of Hole Size	0.7520	0.5028		3.7522
	0.7500	0.5000		3.7500
Limits of Shaft Size	0.7475		5.0000	
	0.7463		4.9990	
Tolerance on Hole				
Tolerance on Shaft		0.0016		0.0014
Allowance			−0.0035	

3. Using the appropriate limit dimensioning tables and either the basic hole or basic shaft method of calculation, complete the portions of the table.

	a.	b.	c.	d.
Basic Size	10	250	25	4
Class of Fit	Loose Running H11/c11	Close Running H8/f7	Locational Clearance H7/h6	Medium Drive H7/s6
Method of Calculation	Preferred Hole Basis	Preferred Hole Basis	Preferred Shaft Basis	Preferred Hole Basis
Limits of Hole Size				
Limits of Shaft Size				
Tolerance on Hole				
Tolerance on Shaft				
Allowance				

	e.	f.	g.	h.
Basic Size	6			50
Class of Fit	Sliding H7/g6	Locational Transition N7/h6	Locational Interference H7/p6	Force U7/h6
Method of Calculation		Preferred Shaft Basis	Preferred Hole Basis	
Limits of Hole Size	6.012		80.030	
	6.000		80.000	
Limits of Shaft Size		160.000		50.000
		159.975		49.984
Tolerance on Hole		0.040		
Tolerance on Shaft	0.008		0.019	0.016
Allowance		−0.052		

CHAPTER

7

COMPUTER-AIDED PRODUCT DESIGN SOFTWARE

☐ INTRODUCTION

Computer-aided Design

Computer-aided design (CAD) is a technology concerned with the use of computer-based tools employed by engineers, architects, and other design professionals in their design activities. CAD is used in the design of such artifacts as products, tools, machinery, buildings, facilities, etc. Current packages range from 2D drafting programs to 3D freeform surface and parametric solid modelers. While 2D drawing remains popular, CAD capabilities have developed well beyond the ability to generate drawings. Today's model-based, object-oriented CAD programs provide designers, engineers, and architects with the ability to digitally capture a product's definition and to integrate this definition with the knowledge-base of the entire enterprise.

Categories of CAD Systems

COMPUTER-AIDED DRAWING

In 1982 Autodesk launched AutoCAD®, the first commercially successful 2D vector-based drafting program. Vector graphics employs geometric elements like points, lines, curves, and polygons to represent images. Since these elements are defined mathematically, they can be stored in a database and later manipulated (e.g., copied, moved, rotated, scaled, arrayed). 2D CAD is used by civil engineers, architects, land developers, interior designers, and other design pro-

fessionals. The principal output of 2D CAD programs are the drawings themselves, rather than a model from which drawings can be extracted. Figure 7-1 shows a 2D CAD drawing created by a student in a first-year engineering graphics course, while Figures 1-14, 1-15, 8-1, and 8-2, appearing elsewhere in this book, are all professionally developed 2D CAD drawings.

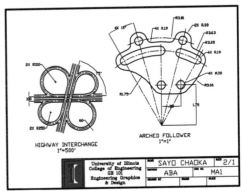

Figure 7-1 2D CAD student drawing (Courtesy of Sayo Chaoka)

Many computer-aided drawing programs also have 3D capabilities. A 3D *wireframe* drawing, like the one shown in Figure 7-2, on page 185 is created using the same geometric elements (e.g., line, circle, arc, polyline) as those used to create a 2D drawing. While only containing edge and vertex information, wireframe drawings provide a relatively simple and fast means to convey the three-dimensional form of an object.

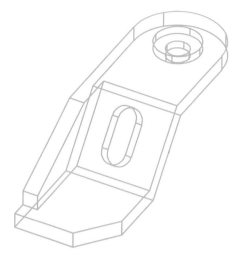

Figure 7-2 3D wireframe drawing

SURFACE MODELING

Freeform surface modeling is used extensively in a number of industries requiring complex, sculpted shapes. Industrial sectors employing sur-

Figure 7-3 Surface model (Courtesy of Fernando Class-Morales)

Figure 7-4 Spline with ducks

face modeling software include automotive, aerospace, shipbuilding, consumer product design, and entertainment animation. Figure 7-3 shows a surface model of a boat created by an undergraduate engineering student.

Surfaces are typically built up from construction curves called **splines**. Historically a spline was a long, smooth, and flexible strip of wood or plastic used on shipyard lofting room floors and in the offices of naval architects to produce the curves that define the shape of a ship's hull. Heavy lead weights called ducks are also used; these maintain the shape of the spline as it passes through fixed data points. The edge of the spline is then used to pass a smooth interpolating curve in either pencil or ink along the edge of the spline. Figure 7-4 shows several ducks and a physical spline positioned to draw a portion of a curve. Traditionally, a **lines plan** describing the geometry of a ship hull, like the one shown in Figure 7-5, would have been created using physical splines.

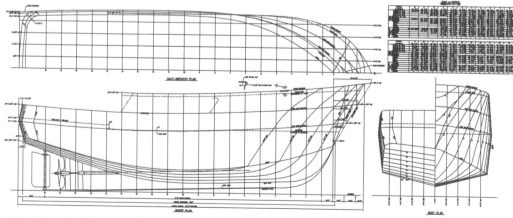

Figure 7-5 Lines plan (Courtesy of Jensen Maritime Consultants, Inc.)

Today, however, these drawings are created using computer software.

The curve resulting from a physical spline can be described mathematically using cubic polynomials. A spline is consequently referred to as a "piecewise parametric polynomial curve." While curves can be represented in either nonparametric or parametric form, the parametric representation of a curve is most useful for the purposes of computer graphics and CAD.

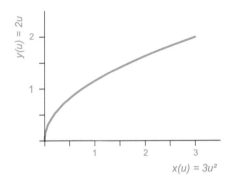

Parametric representation of a curve

In the parametric representation of a curve, each point on the curve is defined by a position vector **P** (see Figure 7-6), where the vector components are:

$$x = x(u), \quad y = y(u), \quad z = z(u)$$
$$\mathbf{P}(u) = [x(u) \quad y(u) \quad z(u)], \quad 0 \leq u \leq 1$$

u	$x(u) = 3u^2$	$y(u) = 2u$
0	0	0
0.1	0.03	0.2
0.2	0.12	0.4
0.3	0.27	0.6
0.4	0.48	0.8
0.5	0.75	1.00
0.6	1.08	1.20
0.7	1.47	1.40
0.8	1.92	1.60
0.9	2.43	1.80
1	3.00	2.00

Figure 7-7 2D parametric curve example

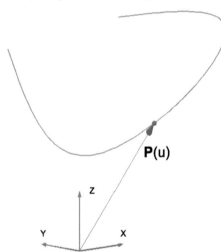

Figure 7-6 Parametric representation of a curve

where x, y, and z are polynomials and u is the parameter.

For example, the coordinates defining the 2D curve shown in Figure 7-7 are generated from the following equations, with the parameter u varying between 0 and 1:

$$x(u) = 3u^2$$
$$y(u) = 2u$$
$$z(u) = 0$$

The underlying mathematics of parametric equations serves as the basis for all freeform curves, including cubic, Bezier, B-spline, and nonuniform rational B-splines (NURBS). Mathematically, a Bezier[1] curve is a special-case B-spline. NURBS, in turn, are a more complex generalization of B-splines. Since NURBS can also be used to represent analytic curves like straight lines, circles, and other conics, most surface modeling programs use NURBS technology to model curves and surfaces.

Figure 7-8 shows an example of a Bezier curve. Bezier curves are characterized by a *control polygon*, which bounds the curve, and *control points*. Note that in the case of a Bezier, the curve passes through the first and

---

[1] Pierre Bezier (1910–1999) was an engineer and creator of the Bezier curves and surfaces used in most CAD and computer graphics systems. Bezier worked at Renault, the French vehicle manufacturer, for many years, where he developed the UNISURF CAD/CAM system.

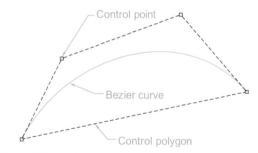

Figure 7-8 **Bezier curve**

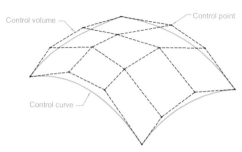

Figure 7-9 **Bezier surface**

last control points, while approximating the interior control points. To modify this curve, the user can simply drag a control point to a new position.

Parametric surfaces are created either directly or by using previously created parametric construction curves. In any case, the mathematics behind the definition of a surface is an extension of spline curves. Figure 7-9 shows a Bezier surface, including the defining curves, control points, and a control volume. Like parametric curves, parametric surfaces can be edited by moving control points. In Figure 7-10A, one half of the symmetrical hull form of a Maine lobster boat is represented by construction curves. Figure 7-10B shows a surface mesh created from these curves, while in Figure 7-10C the surface control points are visible. Figure 7-10D shows a shaded view of the surface.

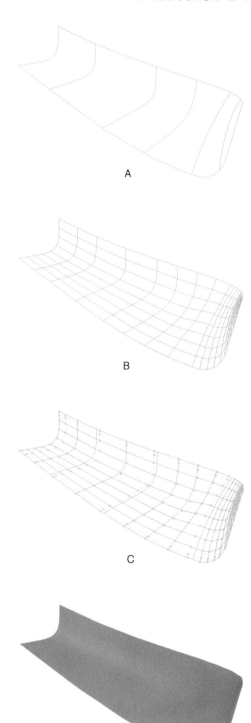

Figure 7-10 Maine lobster boat hull form

Computer-aided industrial design (CAID)

Industrial design is an applied art that strives to improve the aesthetics and usability of products. Computer-aided industrial design (CAID) is a subset of CAD that uses software to assist in the conceptual design of new products. In CAID the focus is on freeform surface modeling. The CAID end product is normally a 3D surface model that can be imported into a solid modeler for development by engineering. CAID software differs from CAD in that it is more conceptual and less technical. See Figure 7-11.

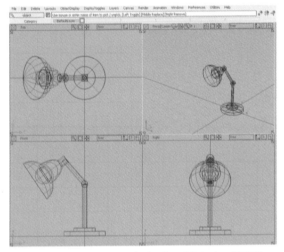

Figure 7-11 Computer-aided industrial design (CAID) software, Alias Studio (Courtesy of Autodesk, Inc.)

Figure 7-12   Solid model created by a student

SOLID MODELING

Figure 7-12 shows an example of a solid model created in AutoCAD® by a student in a first-year engineering graphics course. Notice that, in comparison with the surface model of the speedboat shown in Figure 7-3, the solid model is composed of recognizable solid features like boxes and cylinders.

Solid models are built up from primitives and sweeps. *Primitives* are solid entities similar to the 2D line, circle, and arc entities used in wireframe modeling. Examples of primitives include box, sphere, cylinder, cone, wedge, and torus (see Figure 7-13).

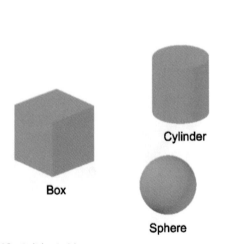

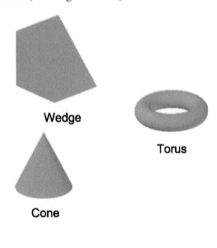

Figure 7-13   Solid primitives

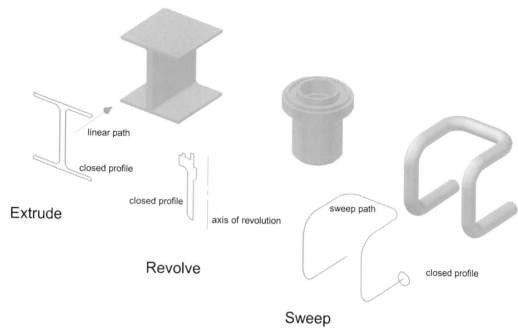

Figure 7-14   Extruded, revolved, and swept solids

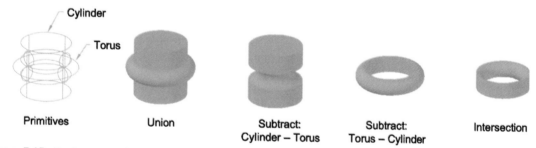

Figure 7-15   Boolean operations

Sweep operations include extruding, revolving, and sweeping. An ***extrusion***[2] is a modeling technique that creates a 3D shape by translating a 2D closed profile along a linear path. A ***revolution*** is a 3D shape formed by revolving a 2D closed profile about an axis. A ***sweep*** is created by moving a 2D closed profile along a path. Examples of extruded, revolved, and swept solids are shown in Figure 7-14. A ***profile*** is simply a 2D outline or shape that does not cross back over

on itself. Most solid modeling operations also require that the profile be closed. Note that all primitives can be created with a single sweeping operation.

The Boolean operations union, subtraction, and intersection are used to combine solid primitives and sweeps. These combined solids are called ***composites***. The results of the different Boolean operations for two different solids are shown in Figure 7-15.

The two most popular representation schemes, or data structures, employed for describing solid models are ***constructive solid geometry (CSG)*** and

---

[2] In manufacturing, an extrusion is a process that involves forcing a material through a shaped opening.

boundary representation (B-rep). The CSG representation stores the model data in terms of the solid primitives and the Boolean operations that are used to combine them. The model history is stored in a ***tree structure***, with the solid primitives serving as ***leaves*** and the Boolean operators as ***branches***. Figure 7-16 shows an example of a CSG tree structure.

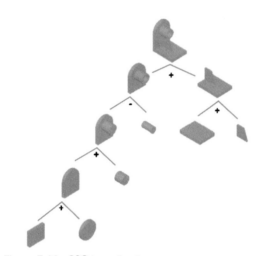

Figure 7-16 CSG tree structure

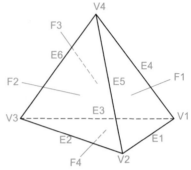

Face Table	
Face	Edges
F1	E1, E4, E5
F2	E2, E5, E6
F3	E3, E6, E4
F4	E1, E2, E3

Edge Table	
Edge	Vertices
E1	V1, V2
E2	V2, V3
E3	V3, V1
E4	V1, V4
E5	V2, V4
E6	V3, V4

Vertex Table	
Vertex	Coordinates
V1	x1, y1, z1
V2	x2, y2, z2
V3	x3, y3, z3
V4	x4, y4, z4

Figure 7-17 Boundary representation (B-rep) of a simple polyhedron

By organizing the data solely in terms of primitives and Boolean operations, CSG data structures are very simple and compact. On the other hand, by only storing the primitives and Boolean operations, information regarding the faces, edges, and vertices of a solid is not readily available. This is a significant shortcoming, since this information is frequently needed when working with solid models. For example, when shading a solid object, information regarding the surfaces of the object is required. Surface information is also needed in preparation for machining a part, in order to calculate numerical control tool paths.

Boundary representation, on the other hand, stores the boundaries of the solid (e.g., vertices, edges, faces) in the database, along with information regarding how these entities are connected. Figure 7-17 shows a simple solid with labeled faces, edges, and vertices. The figure also includes tables that show how the geometry is stored. In B-rep, the ***outside*** of a face is deter-

mined by the order of its bounding edges. When a face is viewed from the outside, the bounding edges of the face are stored in the database in a counterclockwise direction around the face. For example, the edges of face F1 are ordered as follows: E1, E4, and E5. Since the face, edge, and vertex information of a solid is directly available, shading, hidden-line removal and other types of display are more efficient operations using boundary representation.

Most commercial solid modelers employ a combination of both CSG and B-rep for storing models. By adopting this hybrid approach, the strengths of both systems are retained. The drawback to storing models using both representations is, predictably, increased file size.

With solid models like the ones described thus far, it is possible to determine the mass and other material properties of a modeled part. It is also possible to display these solids in various ways, including wireframe, with hidden lines removed, shaded, and rendered. There are, however, a number of capabilities that are not available, among them:

- It is difficult to change geometry. Existing geometry can be added to, subtracted from, intersected with, or the model can be deleted entirely and started over. It is not, however, easy to tweak the dimensions of an existing solid.

- Features only contain geometric information, e.g., a hole is created by subtracting a cylinder. Associated manufacturing and other information is not included in the model database.

- Dimensions are nonparametric: they are added after the primitive is created. The dimensions are not associative with the geometry; they do not *drive* the geometry.

- The model history is unavailable; that is, the order of operations is unavailable to the operator.

- There is no true assembly modeling environment; only parts can be modeled. Without an assembly environment, the motion of moving parts cannot be investigated, and interferences between parts cannot be detected.

In order to emphasize these shortcomings, the models discussed in this section are sometimes referred to in industry as *dumb* solids.

PARAMETRIC MODELING

3D parametric solid modeling addresses all of the shortcomings of dumb solids listed in the previous section. That is, when using a parametric modeler, geometry is easy to modify, features contain manufacturing as well as geometric information, dimensions are parametric, the model history is available to the operator, and there is an assembly environment. So significant, in fact, is parametric modeling that it will be discussed in detail later in this chapter.

The four principal MCAD (mechanical CAD) companies, Autodesk, Dassault Systèmes, Parametric Technology Corporation (PTC), and UGS Corporation, all include at least one parametric modeler in their product lines. The commercial parametric solid modeling software market includes both midrange and high-end packages. The most important midrange packages are Autodesk Inventor, SolidWorks (Dassault), and Solid Edge (UGS). High-end packages include CATIA (Dassault), Pro/ENGINEER (PTC), and NX (UGS). Whereas all of the midrange packages only operate on the Windows platform, high-end packages have offerings that run on UNIX as well as Windows.

The high-end parametric modeling packages tend to be all-in-one solutions, with surface modeling, analysis, manufacturing, collaboration, and other modules all available from the same company. With their ability to work with surfaces and solids (hybrid modeling), high-end solutions allow the designer to incorporate more organic and ergonomic features into their designs.

CAD Viewing and Display

CAD viewing tools like pan, zoom, and orbit are used to control the position from which an object is viewed. When using a viewing tool, it is important to keep in mind that it is the viewer that moves, and not the object or scene.

Display refers to the way in which objects appear on the computer monitor. CAD display types include wireframe, hidden line/hidden surface, shaded, and rendered, as seen in Figure 7-18 on page 192. Using wireframe display, model edges and contours are visible. Wireframe is simpler and faster than either hidden or shaded display, and is frequently used when editing, since all model edges and vertices are available for selection. Solid, surface, and wireframe models can all be displayed in wireframe mode.

Hidden line or hidden surface display refers to a class of computer graphics algorithms used to determine which edges, surfaces, and volumes in a model are visible from a particular viewpoint. Object faces are opaque when displayed using

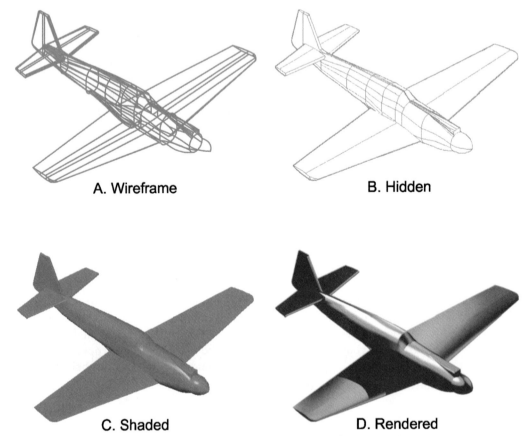

A. Wireframe

B. Hidden

C. Shaded

D. Rendered

Figure 7-18   Wireframe, hidden, shaded, and rendered display (Courtesy of Tim Lingner)

hidden line removal. For this reason, any portion of a face or edge that is obscured by another face from a given viewpoint is not visible. Both solid and surface models can be displayed using hidden line removal.

Shaded display refers to a technique that applies flat colors to visible surfaces. Both solid and surface models can be displayed in shaded mode.

A rendered image is a snapshot of a 3D model to which color, texture, and lighting have been applied to create a single photorealistic image. Both surface and solid models can be rendered. While wireframe, hidden, and shaded displays can all be orbited in real time, rendered views are still images, unless an entire scene is rendered frame-by-frame to create an animation.

☐ PARAMETRIC MODELING SOFTWARE

Introduction

Parametric solid modeling software has the ability to reflect the way in which modern manufacturing companies develop their products. Owing to its object-oriented, parametric nature, parametric modeling has expanded the traditional role of CAD beyond geometry creation and into the realm of product realization.

Unlike primitives-based solid modelers like the one found in AutoCAD®, a parametric modeler provides the operator with multiple work environments, each with its own file type. Work environments common to most parametric modelers include part, assembly, and

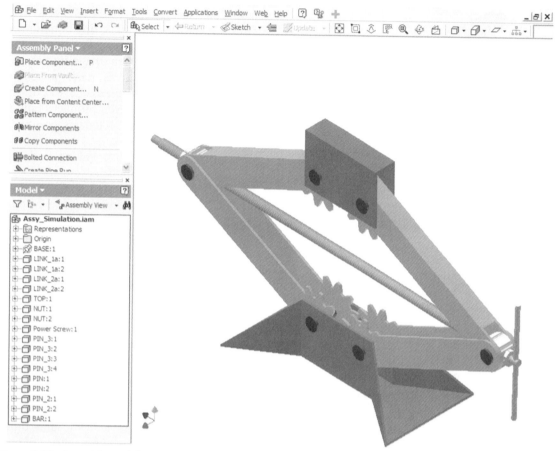

Figure 7-19 Assembly modeling environment (Courtesy of Autodesk, Inc.; Fernando Class-Morales)

drawing.[3] The assembly environment, for example, allows the user to combine components in order to form a virtual product model. Figure 7-19 shows a typical assembly modeling environment.

Most commercial products may be thought of as an assembly composed of different components. A moderately complex product like a bicycle will contain multiple subassemblies as well as individual parts. The rear derailleur, for example, is a subassembly containing many different parts. This natural product decomposition hierarchy is duplicated virtually in the assembly environment tree

structure. Figure 7-20 on page 194 shows the assembly structure for a commercial ball valve.

If the actual product has moving parts, it is possible to constrain the virtual assembly in order to simulate the motion of the part(s). The virtual product model can also be used to identify any static interference between parts, and to determine the range of allowable motion of the product's moving parts.

Within the part environment, this ability to mirror reality continues. Parts are modeled using feature-based techniques, with names like extrude, hole, draft, rib, shell, fillet, and chamfer. These parametric modeling features, as their naming suggests, capture manufacturing as well as geometric data.

---

[3] Examples of other specialized work environments include sheet metal parts and welded assemblies.

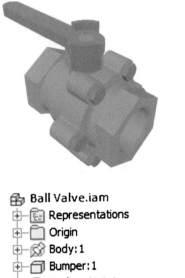

Ball Valve.iam
⊞ 🗋 Representations
⊞ 🗋 Origin
⊞ 🗋 Body: 1
⊞ 🗋 Bumper: 1
⊞ 🗋 Packing Nut: 1
⊞ 🗋 Stem: 1
⊞ 🗋 Ball: 1
⊞ 🗋 Handle: 1
⊞ 🗋 Pin: 1
⊞ 🗋 Body Cap: 1
⊞ 🗋 Component Pattern 2

Figure 7-20   Assembly environment tree structure

Geometric constraints are used to control part geometry by limiting the number of necessary dimensions. These geometric constraints have names like parallel, perpendicular, concentric, collinear, tangent, equal, and symmetric, which reflect the precise language employed by a machinist when fabricating parts.

When creating solid model features, designers will often use work or construction geometry (planes, axes, and points) in order to locate the feature. These work features are analogous to the reference datums employed in manufacturing to fabricate parts. In parametric modeling, dimensions are associatively linked to the model geometry that they describe. If a dimension changes, so does the associated geometry. This gives the user considerable latitude to explore design alternatives. In addition, the parametric basis

of model dimensions allows the designer to build intelligence into the model. Using equations, a parametric dimension can be linked to other parameters of the same feature, different part features, and even different parts within an assembly. This flexibility opens the way to the development of part and even product families, something clearly valued in today's customer-focused, option-driven market.

The concept of associative linking is deeply embedded in parametric modeling system design. Beyond the previously discussed link between a dimension and the size (or location) of a feature, associative links are maintained across the part, drawing, and assembly environments, between an assembly and its associated parts list, between a file's summary information and title block text fields, etc.

Beyond this, other product lifecycle management (PLM) software programs are able to link associatively with the parametric model. As an example, a CAD model can be imported into a finite element analysis (FEA) program for stress analysis. If results indicate an unsatisfactory level of stress, the CAD model can be modified, the geometry updated within the FEA program, and the analysis repeated. This ability to use the CAD model for analysis, manufacturing, and other downstream applications is a cornerstone of product lifecycle management, and is discussed in further detail later in this chapter.

Terminology

Parametric solid modeling employs parametric dimensions and geometric constraints to define part features and to create relationships between these features in order to create intelligent part models. These parts can then be combined to form virtual assembly models. Drawing documents can then be extracted from both part and assembly models.

Features are the basic three-dimensional building blocks for creating parts. Model features available in a parametric modeler mimic actual design and manufacturing features (e.g., counterbore hole, rib, draft angle). There are essentially two kinds of model features, sketched and placed. Sketched features (e.g., extrude, revolve) require that a 2D sketch be made before the feature can be created. Placed features (e.g., hole, chamfer, fillet, shell, face draft) can be created without sketch geometry. Since features are so important to the part modeling process, parametric modeling is sometimes called *feature-based modeling*.

A *parameter* is a named quantity whose value can be changed; for example, d0 = 10. Here d0 is the name of the parameter, 10 is its value. Parametric dimensions control the size and position of features. If the value of a parameter is changed, the feature geometry changes as well. Because of this we say that parametric models are dimensionally driven. In addition, because a parameter has a name as well as a value, parametric relationships can be formed across features, parts, and assemblies. A simple example of this would be d0 = 2*d1. If d1 is 5, then d0 is 10. If d1 is changed, d0 will change by a factor of two.

Constraints are mathematical requirements imposed on the geometry of a 3D model. Regarding part models, there are two kinds of constraints, dimensional and geometric. Dimensional constraints (also called parametric dimensions) place limits on the size or position of a feature. Geometric constraints (e.g., parallelism, tangency, concentricity) place limits on the shape or position of a feature. In some software packages, geometric constraints are called relations. In addition to part constraints, there are also assembly constraints. These constraints determine how different parts are positioned with respect to one another. Examples of assembly constraints include mate and insert. Parametric modeling is sometimes referred to as *constraint-based modeling*.

Part Modeling

INTRODUCTION

The part creation process starts with a two-dimensional sketch. This sketch, normally a closed profile, is used to create the first or *base feature* of the part. The base feature is frequently either an extrusion or a revolved feature. Additional features are then added until the part is complete. These features can either be based on a sketch, or directly placed on the model without the need to create additional geometry.

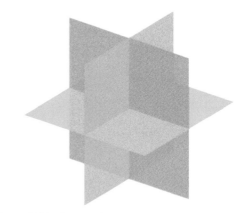

Figure 7-21   Datum planes

New part files contain three mutually perpendicular datum planes similar to those shown in Figure 7-21. One of these planes is selected and used to sketch the profile for the base feature. The origin of the part file is defined at the common intersection of the three reference planes. In some parametric modeling programs three default work axes and a common origin point are also provided, as shown in Figure 7-22.

SKETCH MODE

As seen in the previous section, 2D sketches play an important role in the creation of 3D parametric parts. To support this crucial aspect of part creation, a sketch mode, sometimes called a sketcher, is included within the part modeling environment. Figure 7-23 on page 196 shows a typical sketching interface.

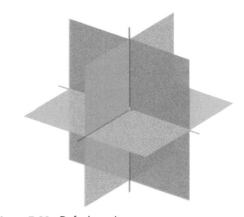

Figure 7-22   Default work geometry

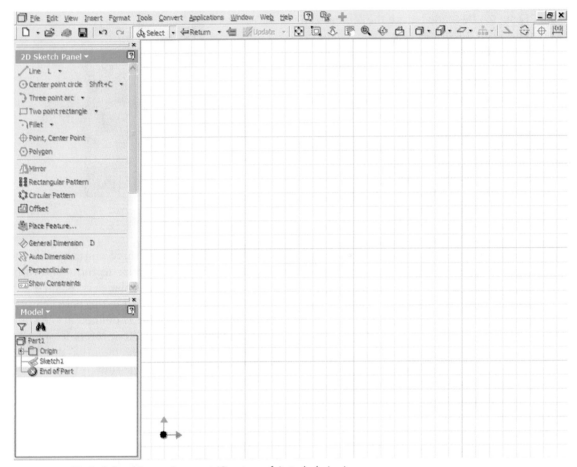

Figure 7-23 Typical sketching environment (Courtesy of Autodesk, Inc.)

Prior to entering the sketching environment, a sketch plane must be defined. This sketch plane can either be 1) one of the default datum planes, 2) a planar face on the existing part geometry, or 3) a new work plane. Figure 7-24 shows examples of these three cases.

Tools found in sketch mode are similar to those found in a 2D CAD program, and include line, circle, arc, trim, offset, etc.

Depending upon how the sketch is made, some geometric constraints may be inferred. Figure 7-25, for example, shows a line being drawn from an arc. When the line appears to be approximately tangent to the arc, a small symbol called a *glyph* appears. If the endpoint of the line is then selected, a tangent constraint is applied between the line and the arc.

After the rough sketch is complete, additional geometric constraints are added. Figure 7-26A shows a recently completed sketch. Figure 7-26B

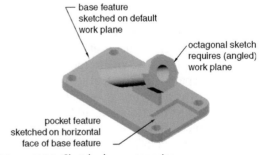

base feature
sketched on default
work plane

octagonal sketch
requires (angled)
work plane

pocket feature
sketched on horizontal
face of base feature

Figure 7-24 Sketch plane categories

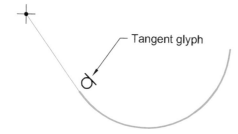

Figure 7-25   Inferred geometric constraints

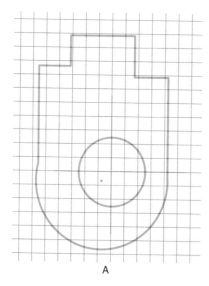

A

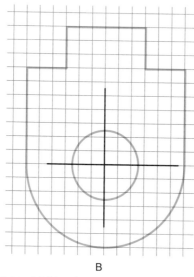

B

Figure 7-26   Additional geometric constraints

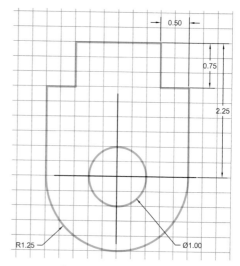

Figure 7-27   A fully constrained sketch

shows the same sketch after concentric, tangent, collinear and symmetric constraints have been applied. Once the sketch has been geometrically constrained, parametric dimensions are added in order fully to constrain the sketch. Figure 7-27 shows a fully constrained sketch.

FEATURE CREATION

Upon exiting sketch mode, the feature creation tools become available. The most commonly used sketched features include extrude, revolve, sweep, and loft (blend). Extrude and revolve both require a single sketch, while sweep and loft need more than one available sketch.

The base feature geometry is always additive. Subsequent sketched features are added, subtracted, or intersected with the existing geometry, depending upon which Boolean operator (join, cut, or intersect) is selected. Figure 7-28 on page 198 shows the result of joining, cutting, and intersecting an extruded sketch with a base feature.

Placed features require that the user select existing geometry and supply specific parametric inputs; no sketches are needed. Figure 7-29 on page 198 shows a part with several placed features, including draft, shell, counterbore hole, and fillet.

Work features (also called construction geometry) include work planes, work axes, and work

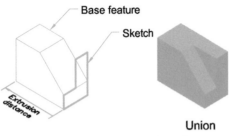

Union   Subtraction   Intersection

Figure 7-28   Boolean operations on a base feature

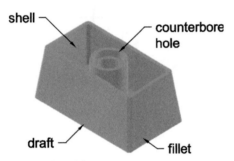

Figure 7-29   Placed features

points. Unlike sketched and placed features, work features do not have a direct effect on the model geometry. Rather, work features are employed to enable the creation of a subsequent geometric feature. For example, both the fuselage and wings of the airplane shown in Figure 7-30 were created using a loft feature. A loft requires multiple sketches. These sketches, in turn, require

sketch planes. To create each loft, multiple work plane features were created by offsetting from a datum plane. Sketches were then made on each of the work planes, and lofted solids were created that pass through the sketches.

In the sidebar a simple procedure describing the part creation process is provided.

Part Creation Process

1. Define a sketch plane
2. Create a rough sketch
   • Some geometric constraints are inferred
3. Add additional geometric constraints
   • Coincident, tangent, symmetric
4. Add dimensions to fully constrain sketch
5. Create the base feature
6. Add additional features
   • Extrude, hole, fillet, chamfer

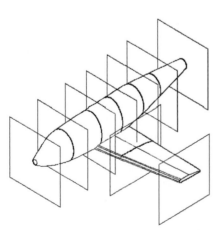

Figure 7-30   Use of work planes to create loft features (Courtesy of Michael Rybalko, Bob Wille, Paul D. Arendt, Stephanie Schachtrup, Dominic Menoni, Daniel J. Weidner)

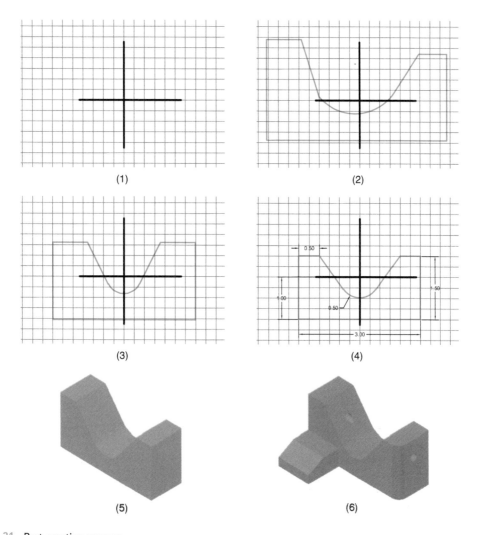

Figure 7-31   Part creation process

PART EDITING

Once created, part features can be modified at any time, either by editing its ***consumed sketch***[4] or by changing inputs and/or parametric values made when the feature was created. In Figure 7-32A on page 200, a part along with its base feature

[4] A sketch is said to be consumed once it has been used to create a feature. In some parametric modeling programs, the consumed sketch can be *shared*, making it available once again for creating additional features.

sketch is shown. Figure 7-32B shows the same part after the base feature sketch has been modified by changing the angle from 50 to 70 degrees. Figure 7-32C shows the part after the extruded distance of the modified base feature has been increased from 60 to 80.

Features may also be ***suppressed***, meaning that the effect of the feature operation upon the part is not applied. The right side of Figure 7-33 on page 200 shows the effect of suppressing hole pattern and fillet features on a garlic press handle. It may be desirable to suppress certain features prior to

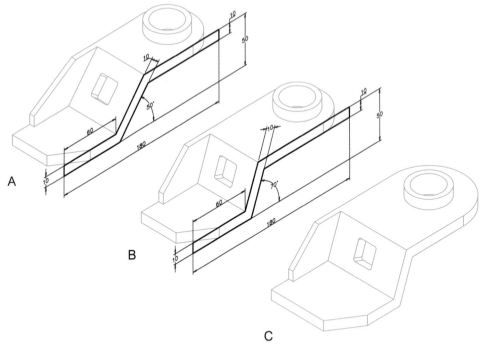

Figure 7-32 Feature editing

importing the part geometry into another PLM software package. For example, small features may be suppressed prior to importing them into a finite element analysis package, since these features may unnecessarily increase the size of the analysis model.

Access to part features, whether to edit, suppress, rename, or otherwise manage them, is easily gained via the *feature tree*. Figure 7-34 shows a simple part and its associated feature tree. Notice that work features (i.e., Work Plane1) appear in the tree structure, as well as sketched and placed features, and unconsumed sketches. Expanding a sketched feature reveals its consumed sketch. As new features are added, they appear at the bottom of the feature tree list.In this way the feature tree captures the model history, showing the order in which the model was created. The

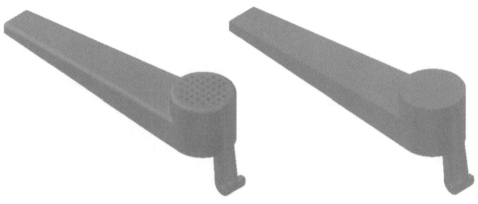

Figure 7-33 Feature suppression

simple part
- ⊞ Origin
- Shared Sketch
- ⊞ Base Cylinder
- ⊞ Minor Cylinder
- ⊞ Notch
- Work Plane 1
- ⊞ Plate
- ⊞ Hole 1
- ⊞ Hole 2
- End of Part

Figure 7-34   Simple part with feature tree

different parametric modeling packages have various tools that allow a user to review the way in which a model has been constructed.

In some modeling packages, a feature's position within the feature tree may be modified by dragging it up or down the tree structure, unless feature dependencies prevent this. These feature dependencies are called **parent-child relationships**. A parent-child relationship refers to the way in which one feature is derived from, and is consequently dependent upon another feature. This dependency is created when a new child feature is positioned with respect to an older parent feature. It is possible to avoid many dependencies by positioning features with respect to the default datum planes, rather than to other features. Note that a child feature must always appear below its parent in the feature tree, and that the base feature will typically be a parent to all other features. If a parent feature is deleted, then any children will be deleted as well. The

raised cylinder shown in Figure 7-35A, on page 202 for example, is the parent to the hole, fillet, and chamfer features shown in Figure 7-35B.

Assembly Modeling

INTRODUCTION

Used extensively in the automotive and aerospace industries, assembly modeling is a relatively new technology that allows the operator to combine components to create a 3D parametric assembly model. Assembly modeling is an essential tool for any work group engaged in the development of a product composed of multiple parts. Some benefits of assembly modeling include:

- A fully associated bill of materials (BOM) can be extracted from the assembly

- The weight, center of gravity, and other inertial properties of the assembly can be automatically tracked

- Interferences between parts can be detected, avoiding embarrassing and costly design mistakes

- Kinematic analysis of moving parts within the assembly can be conducted, including the determination of the range of motion of parts, and the position, velocity, and acceleration of linkages

- Exploded and assembly section views can be easily created (see Chapter 8 for further discussion)

The initial task conducted within the assembly modeling environment is to bring the components into the assembly. Options for bringing a part into an assembly include 1) directly importing the component, 2) creating the part within the assembly environment, and 3) importing a standard part from an internal part library.

Once in the assembly environment, the components must be correctly positioned relative to one another in the assembly. The following two sections discuss how this is accomplished.

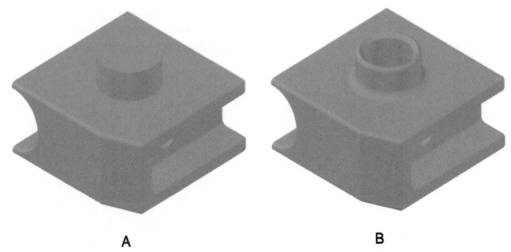

A

B

Figure 7-35 Parent-child relationships

DEGREES OF FREEDOM

Solid part models are rigid bodies and consequently have six degrees of freedom (DOF), three in translation and three in rotation. When a solid part is initially brought into the assembly environment, all six degrees of freedom are open. The part is unconstrained, free to translate along or rotate about any combination of the three coordinate axes. By removing these degrees of freedom, the part's freedom to translate and/or rotate is limited, either partially or completely.

ASSEMBLY CONSTRAINTS

Assembly constraints are parametric constraints that restrict a part's motion relative to another part. This is accomplished by removing degrees of freedom. Typical assembly constraints available in most parametric modeling packages include mate and insert. If a part is fixed with respect to another part, then assembly constraints must be applied to remove all six DOFs. A moving part, on the other hand, is only partially constrained, with unconstrained DOFs in the allowable direction(s) of motion.

Since assembly constraints are parametric, a value is associated with the constraint. This value can represent either a distance or an angle. For example, in Figure 7-36 a mate constraint is applied between the surfaces on two parts. When the parametric offset value for the distance is set to zero (Figure 7-36A), these surfaces are in

contact. If the offset is set to another value, the two surfaces will be offset from one another by that distance (see Figure 7-36B).

CAD LIBRARIES

Most parametric modeling packages include a library of standard parts that is linked to the assembly environment. Standard parts include threaded fasteners, washers, o-rings, etc., as well as structural shapes. Figure 7-37 shows the interface for one of these CAD libraries.

In addition to these internal libraries, external CAD part libraries can be accessed via the Internet.

Advanced Modeling Strategies

What is the optimal modeling approach, anticipating that the model will need to be modified in the future? This question captures the essence of *design intent*, a term frequently used in conjunction with parametric

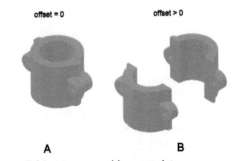

offset = 0   offset > 0

A   B

Figure 7-36 Mate assembly constraint

Figure 7-37 Standard part library interface (Courtesy of Autodesk, Inc.)

modeling. While parametric features are easily adjusted, a parametric model built from these features may be easy, difficult, or nearly impossible to modify. This is a result of the feature dependencies that tend to accumulate as a model is built.

The goal of design intent is to optimize the usability of the model. To achieve this, operators must carefully weigh the consequences of their modeling decisions. The easy solution today may unfortunately make the model unusable in the future. Careful planning is necessary if a parametric design is to function as intended.

As an example, assume that equally spaced holes are to be made in a flat bar. Given the following dimensions:

• Flat bar length (L) = 48″

• Number of holes (N) = 5

• Distance from flat bar ends to hole center nearest to the end (D) = 2″

this model can be built in three steps:

1. Create the flat bar

2. Create a hole at one end of the bar

3. Use a rectangular array to pattern the other holes. Note that the hole spacing must be determined.

The result is shown in Figure 7-38A on page 204. Now assume that a manufacturing company uses a family of parts like the one described previously, where L, N, and D are the driving parameters. In other words, the design intent is to be able to enter a flat bar length, the number of holes, and the distances from the ends of the flat bar to the nearest hole, and arrive at the desired part. Figure 7-38B shows the same part, but with L = 36″, N = 8, and D = 1″. Figure 7-38C shows the parameters dialog box used to establish the relationships that allow the part to be updated as intended, by simply changing the three parameters.

Figure 7-39A on page 204 provides another example. A manufacturing company uses wire shelving similar to that shown in the figure. Longitudinal and transverse lengths of wire are welded together to form the shelf. Relevant dimensions include the diameter and length of the wire, the number of wire lengths used, the spacing between the lengths of wire, and the overhang, for both the longitudinal and the transverse wires. It is fairly easy to create this model as a single part, for example:

1. Create a longitudinal wire

2. Array

3. Create a transverse wire, making sure that this wire just touches the longitudinals.

4. Array

However, if the intent is to vary these parameters (diameter, length, count, spacing, and overhang), in order to generate a family of wire shelving configurations (see Figure 7-39B), then a good deal of planning is necessary.

Parametric modeling can also be used to create entire product families. Figure 7-40 on page 205 shows a popular mixer blender design employed in the food processing industry. The mixer includes multiple subassemblies, and contains several hundred parts. This mixer is available in a wide range of capacities. In addition, customers require that the product be customized to suit the particular needs of their facilities, which can impact, for example, the parameters shown in Figure 7-41 on page 205.

A senior design project team was faced with the problem of developing a product family model of the mixer blender for an industrial sponsor. The team first modeled a specific capacity mixer with the intent to use it as the basis for a product family. As suggested in Figure 7-42, the design team then linked

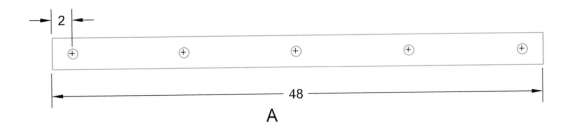

A

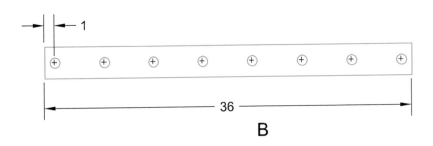

B

Parameter Name	Units	Equation	Parameter Value
L	in	36 in	36.00000
D	in	1 in	1.00000
N	-	8	8.00000
hole_spacing	in	(L - 2*D)/(N - 1)	4.8571

C

Figure 7-38 Plate with holes part family

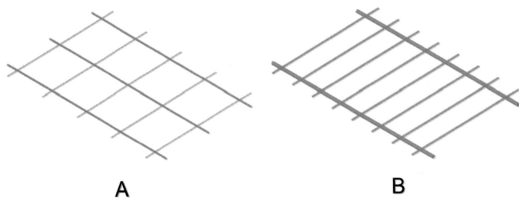

A   B

Figure 7-39 Wire shelf part family

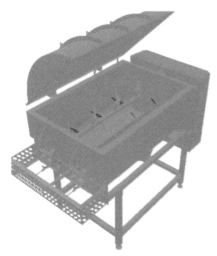

Figure 7-40 Mixer product family (Courtesy of Cozzini Inc.; Adam R. Andrea, Katie Kopren, Philip Kunz)

the model assembly parameters via multiple spreadsheets to successfully create the product family model. After specifying the driving parameters in a master spreadsheet, the sponsor was able to generate a mixer assembly model of the specified capacity.

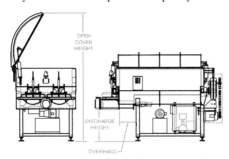

Figure 7-41 Mixer client specified dimensions (Courtesy of Cozzini Inc.; Adam R. Andrea, Katie Kopren, Philip Kunz)

Building information modeling

Although the architecture, engineering, and construction (AEC) industry has lagged behind mechanical CAD in terms of adopting parametric, model-based software solutions, this now appears to be changing. ***Building information modeling*** (BIM) has been succinctly described as "3D, object-oriented, AEC-specific CAD." The American Institute of Architects has further defined BIM as "a model-based technology linked with a database of project information." BIM provides a digital representation of the building process to facilitate

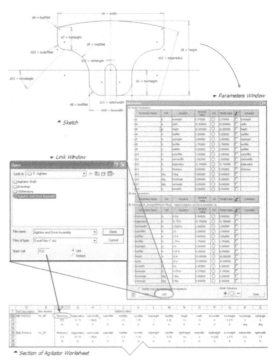

Figure 7-42 Mixer product parameter roadmap (Courtesy of Cozzini Inc.; Katie Kopren, Philip Kunz)

compatibility and exchange of information in digital format, and reflects the general reliance on database technology as the foundation. Both Autodesk and MicroStation market BIM software solutions. Figure 7-43 shows an image from the Autodesk Revit BIM environment.

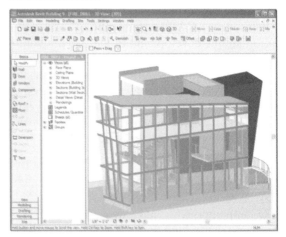

Figure 7-43 Building Information Modeling interface (Courtesy of Autodesk, Inc.)

☐ DOWNSTREAM APPLICATIONS: CAD/CAM/CAE INTEGRATION

Along with design, other key elements of the product development process include manufacturing and engineering analysis. While computer-aided manufacturing (CAM) focuses on product manufacture, computer-aided engineering analysis (CAE) concentrates on product analysis. Although developed in isolation, CAD, CAM, and CAE are all computer-based technologies focusing on product definition. It is consequently not surprising that the integration of these technologies is an important goal of companies engaged in product development.

Computer-aided manufacturing is the technology concerned with the use of computer systems to plan, manage, and control manufacturing operations. The most important CAM application areas include numerical control (NC), robotic control, and computer-aided process planning (CAPP). NC is the technique of using programmed instructions to control a machine tool that grinds, cuts, mills, punches, bends, or turns raw stock into a finished part. Robotic control refers to the use of robots to select and position tools and work pieces for NC machines. Using engineering drawing data as input, process planning is the act of preparing detailed work instructions that describe the type and sequence of manufacturing processes to be used to produce a finished part. CAPP systems employ software to automate the development of these manufacturing process plans.

Computer-aided engineering analysis is the application of computer software to analyze geometry, allowing the designer to simulate and study how a product behaves in order to refine and optimize the design. CAE application areas include:

- Finite element analysis (FEA)—structural and thermal analysis of components and assemblies
- Computational fluid dynamics (CFD)—thermal and fluid flow analysis
- Kinematics—determine motion paths and linkage velocities in mechanisms

- Large displacement analysis programs—determine loads and displacements in complex assemblies such as automobiles
- Optimization and design automation tools
- Manufacturing analysis tools—simulate and study manufacturing processes (e.g., casting, molding, die press forming)

In this decade some interesting trends involving CAD/CAE integration have emerged. In a number of instances, important CAD companies have purchased CAE software companies.[5] These purchases have resulted in expanded capabilities of midrange CAD companies, which now offer embedded CAE modules capable of FEA and motion analysis. Another discernible trend concerns the licensing of CAE software.[6] In this arrangement, the customer purchases a license for the CAE software, and also specifies a CAD application. The CAD files can then be opened directly in the CAE application, with associative linking between the two. Figure 7-44 shows screen captures taken from the ANSYS Workbench FEA package. The model itself was created in an associatively linked CAD package.

As manufacturing companies began to implement CAD, CAM, and CAE technologies, the management of product data became a significant problem. In response, software companies started to develop solutions to address the problem. This effort resulted in the emergence of *product data management (PDM)*, a category of computer software that links product data within a database. The information being stored and managed includes such engineering data as CAD models and drawings,

---

[5] Dassault Système's purchase of Structural Research & Analysis Corporation (SRAC) and its COSMOS analysis software in 2001; Autodesk's purchase of Solid Dynamics in 2005.

[6] Examples include ANSYS Workbench, MSC.Software's Dynamic Designer, and Blue Ridge Numerics CFDesign software.

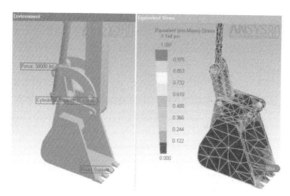

Figure 7-44 ANSYS Workbench screen captures
(Courtesy of ANSYS, Inc.)

and their associated documents. PDM software capabilities include:

- Product data storage in a secure, centralized vault. This data includes:
  - CAD files
  - Third party CAD files
  - Non-CAD files
- The ability to assign and control access to the data
- Search tools to locate and preview the data
- File management tools for check-in, check-out, and document release
- Version management tools to avoid unintentional overwriting of files
- Viewing and markup tools

In the quest to further integrate CAD, CAM, CAE, and PDM software, the term **product lifecycle management** (**PLM**) has been coined. While PLM as a discipline emerged from these software technologies, PLM can be understood as the integration of these tools with methods, people, and processes through all stages of a product's life. It is a framework that encompasses and integrates existing design, manufacturing, and communication technologies.

The PLM model has not been universally adopted within the MCAD community. Autodesk, for instance, is promoting **digital prototyping** as a construct for understanding CAD/CAM/

CAE integration. Digital prototyping is the *digital simulation* of a product in order to test form, fit, and function. The goal of digital prototyping is to capture all of the geometric and functional characteristics of the product in a single model, and to replace physical with digital prototyping, testing, and simulation.

While the PLM umbrella is more focused on the entire product lifecycle, digital prototyping has more of a design focus. The technology, both hardware[7] and software, has matured to the point where industrial designers can now capture their ideas in digital format, thus expanding the digital pipeline to include concept design. Engineers and industrial designers are now able to work collaboratively in the same digital medium.

Another significant development associated with digital prototyping is *functional design*. Functional design software allows engineers to concentrate on the function of a design, rather than on its geometric form. In functional design mechanisms (e.g., gears, shafts, springs, belts and pulleys, bolted connections, cams) are designed by specifying the key functional parameters that drive the design. Once completed, the parametric geometry is automatically created. The bevel gears shown in Figure 7-45, for example, were

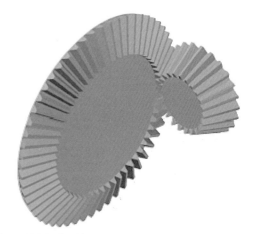

Figure 7-45 Bevel gears created using functional design
(Courtesy of Autodesk, Inc.)

[7] The Cintiq interactive pen display from Wacom, for example, allows designers to sketch directly on the monitor screen.

created using a dialog box to input the relevant parameters. No sketches were made, features created, etc. Functional design software also includes analysis capabilities.

□ DOWNSTREAM APPLICATIONS: RAPID PROTOTYPING

Introduction

The term *rapid prototyping* refers to the automatic construction of physical objects directly from a geometric solid or surface model. While several different rapid prototyping (RP) technologies have emerged since the 1980's, all of them start with a virtual CAD model. The model geometry is then transformed into cross sections, and the prototype is additively fabricated one physical cross section at a time.

Unlike part production using numerical control (NC) machine tools, rapid prototyping is fast and simple. Rapid prototyping does not require process planning, tooling, or material handling. While NC machine tools can work with most materials, including metals, rapid prototyping is currently restricted to the use of specific materials (e.g., ABS plastic). For this reason, RP physical objects are often used as prototypes or patterns for other manufacturing procedures.

Rapid prototyping is used for 1) design evaluation, 2) function verification, 3) aiding in the creation of models for other manufacturing processes, and 4) producing construction quality parts in relatively small numbers. A physical model, particularly one that can be quickly generated, allows all parties involved in the development process to visualize, discuss, and intelligently evaluate a particular design. In this way, potential problems and misunderstandings are uncovered and resolved, thus avoiding costly mistakes that go undetected until late in the product's development cycle.

Rapid prototypes are also used to verify that a design will function as intended. Common function verification tasks include:

- Demonstrating the practicality of an assembly. Many products are difficult, or even impossible to assemble.

- Evaluating the kinematic performance of an assembly. Do moving parts perform as intended? Are there any unexpected interferences?

- Assessing aerodynamic performance. Here the geometric shape is of primary importance; a prototype made from a different material may be sufficient.

In the event that such characteristics as the strength, fatigue, operational temperature limits, or corrosion resistance of a part are to be tested, the prototype must then be made of the same material as the actual part. In this case, RP prototypes are sometimes used as patterns for other fabrication processes.

The most significant benefits of RP technology are a compressed design cycle and improved product quality. While dramatically reducing the time and expense required to take a new product from initial concept to final production, rapid prototyping is also helpful in identifying design flaws.

Technology Overview

Currently the most commercially successful RP technologies include stereolithography, fused deposition modeling (FDM), and selective laser sintering (SLS).

Stereolithography, developed by 3D Systems, Inc., remains one of the most popular RP methods. In fact, the STL file format, originally developed by 3D Systems, is the industry standard for interfacing a virtual solid model with a rapid prototyping machine. A stereolithography apparatus (SLA) builds plastic parts one layer at a time by tracing a laser beam on the surface of a vat of liquid photopolymer. Inside the vat is a table that can move up and down. After each layer is completely traced, the work table is lowered exactly one layer thickness into the vat of liquid, allowing fresh polymer to flow in and cover earlier work. Stereolithography requires a support structure when the part being built has undercuts; that is, when the upper cross section has a larger area than a lower cross section.

Fused deposition modeling (FDM) is most closely associated with Stratasys, Inc. In FDM, each layer is generated by extruding a thermo-plastic material in a liquid state. As in stereo-lithography, parts are built on a platform or table that moves up and down. Once a complete layer is deposited, the table drops down and another layer begins. FDM prototypes also require a support structure.

Selective laser sintering (SLS) employs a pow-der material that is selectively fused by a laser. After a roller spreads the powder, the laser is used to solidify a single layer of material. This process repeats until the part is complete. In SLS, a support structure is not needed, since the unpro-cessed powder serves this purpose. Parts can be produced from a wide range of commercially available powder materials.

STL Files

The de facto standard format used to pass model geometry, either solid or surface, to an RP device is the stereolithography, or STL file. Virtually all CAD programs can save directly to the STL file format. In an STL file, the external boundary of an object is represented by a mesh of triangles, called a tessellated or faceted object. The vertices of each triangle are ordered so that in traversing the vertices in a counterclockwise direction, the surface normal defined by this procedure identifies the outside of the solid.

The simplicity of the STL file format, as a list of triangles, makes file conversion to and from STL particularly easy. Many standard surface triangulation algorithms have been developed in computer graphics, and are used to assist in developing file translation routines. In addition, the accuracy to which the STL file represents the actual geometry can be controlled by increasing the number of triangles. Any 3D form can be represented in STL, and a simple slicing algor-ithm is available to convert the mesh of triangles into cross-sectional slices.

Disadvantages of the STL format include data redundancy, errors due to approximation, and information loss. In an STL file each facet normal is explicitly stored, despite the fact that this infor-mation can be determined from the vertex order. In addition, vertices are stored multiple times,

one for each facet. Since they can only be represented as planar triangles, STL files do a poor job in representing curved surfaces. Finally, in converting from a geometric model to STL format, information regarding the geometry, topography, and material of the part model is lost.

STL files can be saved in either ASCII or binary form. In saving to the STL format, the user can typically control the accuracy of the representation by specifying an increase in the total number of triangles used. The trade-off for this increase in accuracy is increased file size.

Characteristics of RP Systems

Important characteristics common to most RP systems include part orientation, support struc-ture, and hatch style.

PART ORIENTATION

Prior to converting the STL file into layered cross sections, the user should decide how to orient the model. Part orientation can affect build time, part resolution, and surface finish. Higher resolutions of curved surfaces can be obtained by choosing the orientation so that the object's most significant curvature appears in the individual layers, or slices.

As an example, consider a solid cylinder. If the cylinder is oriented so that the axis of the cylinder is normal to the build platform (see Figure 7-46A on page 210), then all of the cross sections will be circles, and the curved surface of the resulting prototype will be very smooth. If however, the cylinder is oriented so that the axis is parallel to the build platform (see Figure 7-46B), the result-ing cross sections will be rectangular, and the curved surface of the cylinder will have a stair-step appearance.

SUPPORT STRUCTURE

Many RP technologies require additional support structures in order to successfully build parts. These support structures are, generally speaking, automatically generated by the software. Support structures are required both to assist in the easy removal of the part from the base platform, and when islands or cantilevered sections exist. An island is a portion of a layer that is not connected

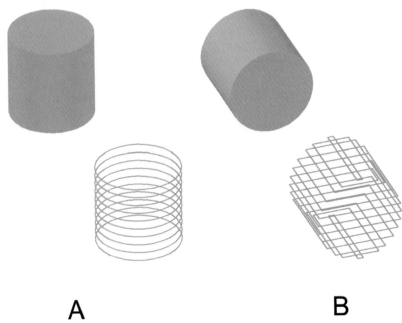

A   **B**

Figure 7-46 Cylinder cross sections: circular versus rectangular

to any other portion of the same layer. Examples of cantilevered features include overhangs and arches. In both instances, upper cross sections have a larger area than lower cross sections. Figure 7-47 on the left shows a printed part prior to removing the support structure; on the right the same part is shown after the support structure has been removed.

HATCH STYLE

In most RP technologies, the user has at least some limited control over the type of internal hatching employed on the interior of solids. For example, a sparse interior can be specified, where solid boundaries are first created, and then a honeycomb interior. Sparse models use less material, reduce fabrication time, and are lighter

in weight than prototypes made with a solid interior. Figure 7-48 shows the internal structure of a block printed with a sparse fill.

3D Printing

While rapid prototyping has certainly had a significant impact on the way modern products are developed, the technology does have some

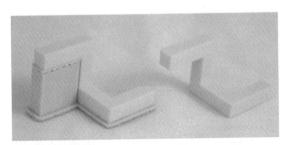

Figure 7-47 Support structure

Figure 7-48 Sparse fill

drawbacks. First, RP technology is simply too expensive for most midsize design offices and product development firms to justify. Starting at about $100k, a rapid prototyping machine can cost as much as $500k. In addition, RP devices are not intended for operation in an office environment. In response to these drawbacks, a number of RP service bureaus have appeared in recent years, and can provide customers with physical prototypes within a short turnaround time.

Another alternative, particularly attractive with design offices and in educational institutions, are 3D printers. 3D printing is a less expensive form of rapid prototyping that can be conducted in an office environment. 3D printer technology offers a favorable price-to-performance ratio when compared to the rest of the RP market. The cost of a 3D printer currently ranges from $15k to $60k, and prices continue to decline. 3D printer sales represent the largest growth segment within the RP market, and this trend should continue. In addition, 3D printers offer much of the functionality of RP technology.

Although RP devices on the whole produce more accurate parts than do their 3D printer counterparts, the differences are not dramatic. RP machines also have slightly larger build volumes[8] than do 3D printers. Further, the software provided with RP machines gives the user more control over the output than does 3D printer software. Perhaps the most significant advantage of RP devices is the ability to use different materials. By and large, 3D printers can only work with a single material.

While more complicated than a conventional printer, 3D printers do not require extensive training in order to operate. By not requiring a special operating environment, 3D printers are gradually coming to be viewed as standard design office equipment, perhaps comparable to a plotter or large photocopier. 3D printers are actually faster than most RP devices. A physical prototype can typically be ready in a day, or less.

Like other RP equipment, 3D printers are used both for design evaluation (visualization, communication, collaboration), and for function verification

Figure 7-49 Z Corp 3D printer (Courtesy of Z Corporation/ZPrinter® 450)

(form, fit, validation, interference detection). To a lesser extent, 3D printers are used either as functional prototypes, or as a model for another manufacturing process.

Three companies hold a significant portion of the 3D printer market, these being Z Corporation, Stratasys, and 3D Systems. Z Corp currently markets three different 3D printer models. All of them employ an inkjet print head that deposits a liquid binder onto a powder. The powder is a low-cost starch or plaster-based material. Z Corp 3D printers (see Figure 7-49) are probably the fastest on the market, and their powder material costs are significantly lower than those of their competitors. While 3D prints made with a Z Corp apparatus have a rough, porous surface, the parts can be impregnated with an epoxy-like resin to achieve a smooth finish and improved strength. Another drawback is, as a powder-based system, there is a lot of dust. The least expensive Z Corp model is monochromatic, but a more expensive model has full-color capabilities.

Stratasys markets two versions of its Dimension 3D printer. The Dimension is based on FDM technology, which employs a moving head that

---

[8] Build volume refers to the largest possible single object that can be made with a RP device. A typical 3D printer build volume is $8'' \times 8'' \times 12''$.

Figure 7-50   Dimension 3D printer (Courtesy of Stratasys, Inc.)

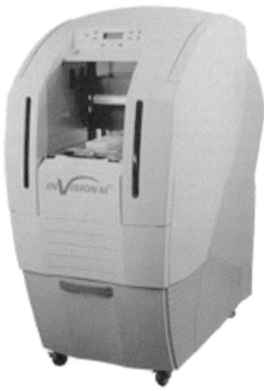

Figure 7-51   InVision 3D printer (Courtesy of 3D systems Corp., InVision® 3-D Modeler)

its parts, these machines are a good choice when testing for form, fit, and function is an issue. The ABS filament is available in several different colors. One version of the Dimension uses a break-away support structure which can involve considerable post-processing time. In the more expensive version, the support structure is water soluble, and is removed by placing the part in an agitation bath. A Dimension 3D printer is shown in Figure 7-50.

Another significant competitor in the 3D printer market is 3D Systems, the first commercially successful RP company. The InVision line of 3D printers has been marketed by 3D Systems since 2005. The InVision (see Figure 7-51) uses an inkjet type print head to deposit proprietary photopolymer materials to form the layers. The InVision employs a melt-away support structure, and is known for producing parts with a good surface finish.

melts and deposits a filament of ABS plastic in order to form the layers. The layers are deposited on a platform that moves down after each layer is completed. The ABS material has properties comparable to many other commonly used plastics. Parts made with these machines are known for their strength and durability. Since the Stratasys 3D printer is also known for the accuracy of

□ QUESTIONS

1. Given the dimensioned isometric view (P7-1 through P7-20), create a solid model of the object.

2. Given the dimensioned assembly view (P8-1 through P8-5), create a solid assembly model.

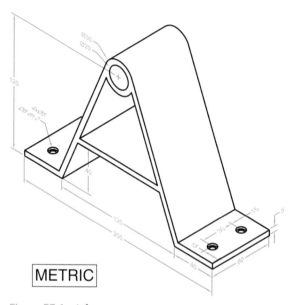

METRIC

Figure P7-1   A-frame

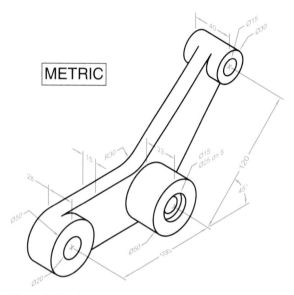

METRIC

Figure P7-2   Arm

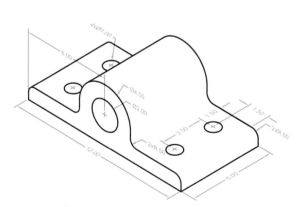

Figure P7-3   Bearing block

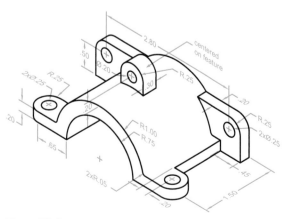

Figure P7-4   Bearing cap

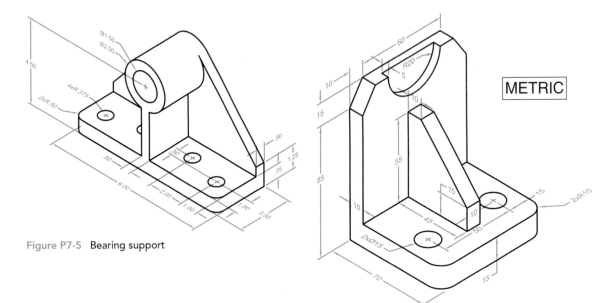

Figure P7-5   Bearing support

METRIC

Figure P7-6   Bookend

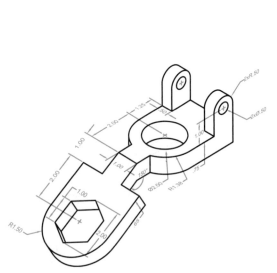

Figure P7-7   Bracket

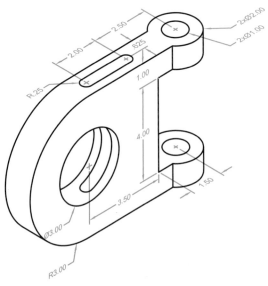

Figure P7-8   Clamp

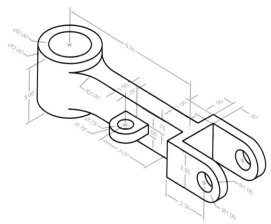

Figure P7-9   Connecting rod

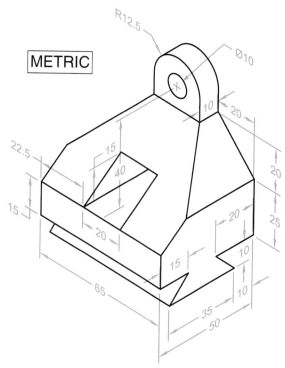

METRIC

Figure P7-10   Dovetailed block

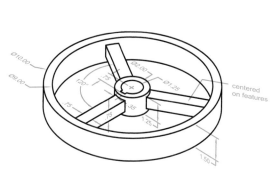

Figure P7-11   Flywheel

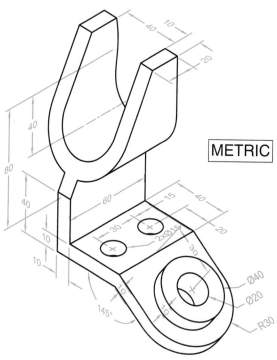

METRIC

Figure P7-12   Fork

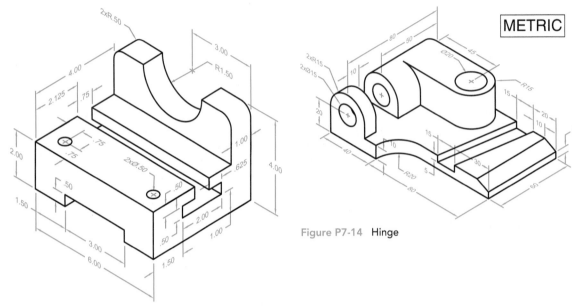

Figure P7-13   Guide

Figure P7-14   Hinge

METRIC

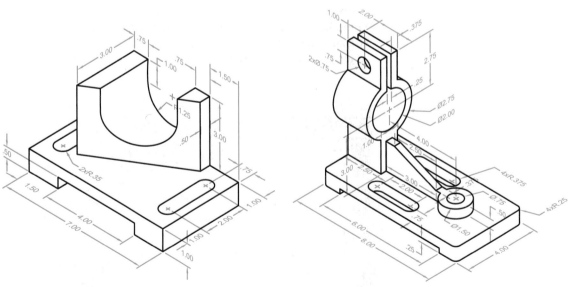

Figure P7-15   Holder

Figure P7-16   Mounting

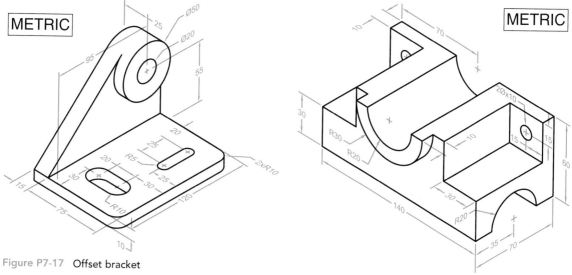

Figure P7-17 Offset bracket

Figure P7-18 Saddle

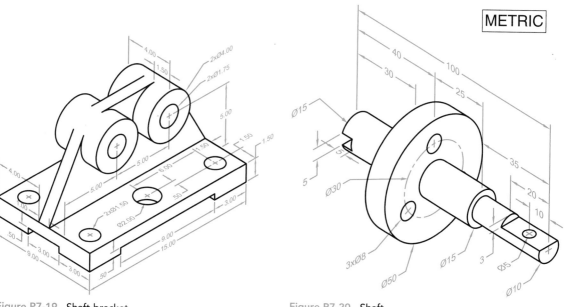

Figure P7-19 Shaft bracket

Figure P7-20 Shaft

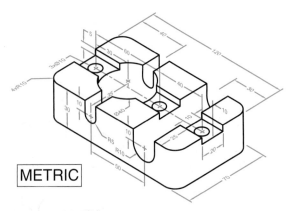

METRIC

Figure P7-21   Slide

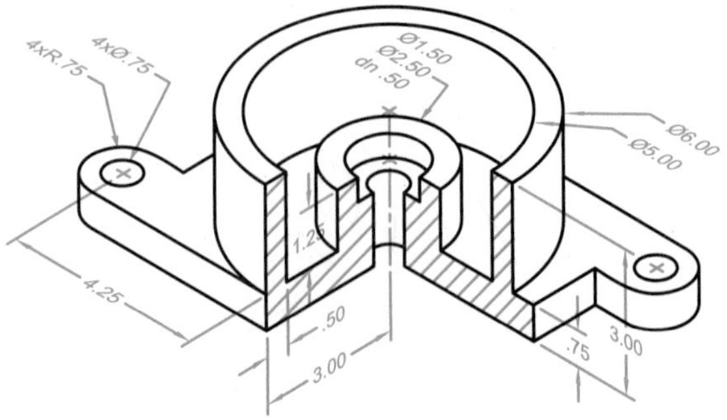

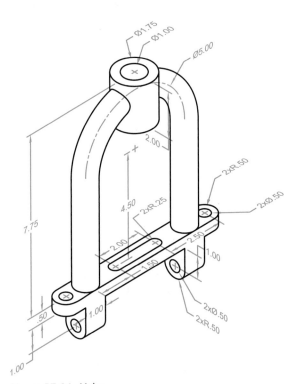

Figure P7-22   Support base

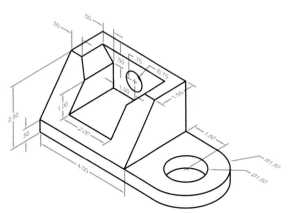

Figure P7-23   Wedge

Figure P7-24   Yoke

8

WORKING DRAWINGS

☐ INTRODUCTION

At the beginning of any project, there is no guarantee that a design will actually be executed. If a project is put out to bid, for example, only one of several competing preliminary designs will be selected. Even then, funding for the project may fall through. Similarly, in a company engaged in both research and development, management may decide to abandon the development of a new product.

Once the decision is made to build, though, the existing preliminary design must be further developed and detailed for production. The term *working drawing* is used to describe the complete set of drawing information needed for the manufacture and assembly of a product based on its design.

As discussed in the previous chapter, commercial products are almost always assemblies comprised of several different parts. Perhaps the most recognizable element of a working drawing set is the *assembly drawing*. The purpose of the assembly drawing is to show how the different components fit together to form the product. An example of an assembly drawing is shown in Figure 8-1.

An essential element of a working drawing is the parts list, or *bill of materials* (abbreviated BOM). The purpose of the BOM is to identify all parts, both standard and non-standard, used in an assembly. The BOM for the assembly shown in Figure 8-1 is located in the upper right corner of the drawing.

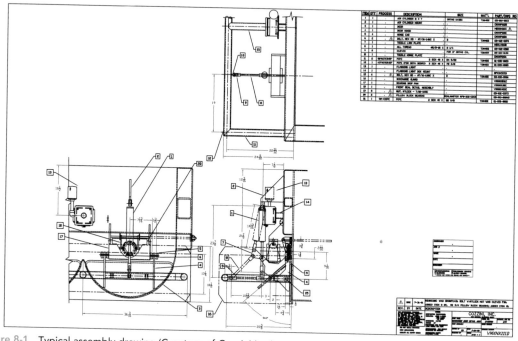

Figure 8-1 Typical assembly drawing (Courtesy of Cozzini Inc.)

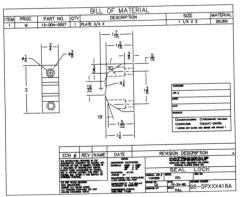

Figure 8-2 Typical detail drawing (Courtesy of Cozzini Inc.)

EXPIRES   07/18/2008

NOT VALID WITHOUT SIGNATURE.

Figure 8-3 Professional engineer's stamp (Courtesy of Jensen Maritime Consultants, Inc.)

In addition to the assembly drawing, a set of working drawings also includes *detail drawings* of all non-standard parts. As seen in Figure 8-2, these individual part drawings contain multiple views, dimensions, notes, tolerance information, material specifications, and any other information necessary to manufacture the part.

Working drawings are typically accompanied by written instructions, called *specifications*, which serve to clarify further the details for manufacturing the product. An excerpt from the specifications describing the construction of a tug boat appears in Table 1-4 on page 15 in Chapter 1.

In addition to describing the details of a product's design and manufacture, working drawings and written specifications also serve as legal contracts. In the event that a problem arises with a design, working drawings may be called upon to help establish liability. Figure 8-3 shows a professional engineer's stamp taken from an engineering drawing. In stamping and signing the drawing, the engineer takes responsibility for the accuracy of the drawing's contents.

☐ THE IMPACT OF TECHNOLOGY ON WORKING DRAWINGS

As CAD technology has matured in the past 25 years, the elimination of the need for engineering drawings has frequently been predicted. While their elimination is perhaps unrealistic, the importance of drawings, at least within some

disciplines, has certainly diminished. In mechanical CAD (MCAD) for example, with its embrace of parametric solid modeling, the digital model has assumed the primary role in product definition, replacing the drawing. Later in this chapter we will see that it is fairly simple to extract 2D drawings from a 3D constraint-based model, either part or assembly. In the architecture, engineering, and construction (AEC) industry, recent strides have been made in the development of building information modeling (BIM) software, showing that this ability to extract drawing information from a 3D model is not unique to the mechanical or aerospace industries.

ASME Y14.41-2003, Digital Product Definition Practices, provides standards for the use of digital data sets as a substitute for working drawings. According to ASME Y14.41-2003, a product data definition set is a collection of one or more computer files that discloses, by means of graphic or textual presentations, the physical or functional requirements of an item. The Y14.41 Standard supports two alternative methods for the development of the product definition data: 1) model only, or 2) model and drawing in digital format.

CAD technology has impacted the way working drawings are used by replacing traditional file storage, with its attendant risks[1] and demands for space, with electronic file storage. Computer networks and associated software allow authorized individuals to view the latest version of a working drawing at any time.

□ DETAIL DRAWINGS

A set of working drawings includes detail drawings of all non-standard parts included in the product or assembly. Standard parts, either vendor-purchased or developed within the company, do not require individual part drawings. A detail drawing is a fully dimensioned multiview drawing that contains all of the information necessary to manufacture the part. Figure 8-4 on page 222 shows a detail drawing of a shaft part.

Part drawings are either developed directly in 2D, or extracted from a 3D model. A typical part drawing includes the following information:

- drawing views
- dimensions
- tolerances
- material designation
- surface finish
- notes
- title block
- revision block

Although many parts are simple enough to be included together with other parts on the same drawing, there are good reasons to place each part on its own drawing. In doing so, a single identification number can be used to identify both the part and the drawing. This same number can also be used to name the CAD file that documents the part. This practice considerably simplifies the need to keep track of an ever-increasing number of parts, drawings, and files that even a small company must manage. In addition, it also makes it easy for the company to reuse parts in different assemblies.

In many companies, parts are identified by a unique code number. This code or identification number contains information about the part. Formal classification and coding techniques have been developed and are used to group similar parts based on such characteristics as material, size, shape, function, process or other information.

□ ASSEMBLY DRAWING VIEWS

The main purpose of an assembly drawing is to show all of the different parts in the assembly, and how these parts fit together to create the mechanism, device, component, or product. Using parametric modeling software, it is relatively easy to generate the views needed to accomplish this task. Two of the most commonly used assembly views are the section and the exploded view.

A section view of an assembly is used when the relationship between the different parts may not be apparent from an external view, as is the case with the ball valve shown in Figure 8-5 on page 223. In addition to the full section, half section and broken out assembly views are also used, as seen in Figures 8-6 on page 223 and 8-7 on page 224 respectively, for a globe valve.

Certain conventions regarding section lines are to be followed when creating assembly section views. These include the following:

- The section lining used for adjacent parts through which the cutting plane passes should be different. Either use different section line patterns for each material (as employed in Figure 8-5), or use the general-purpose (cast iron) section lining and vary the angle to help distinguish different parts (the approach used in Figures 8-6 and 8-7).

- Standard parts and solid parts with no internal detail are not sectioned. These parts include

---

[1] A fire destroyed all of drawing files belonging to an engineering firm the primary author worked for, Nickum, Spaulding & Associates, of Seattle, Washington. In another war story, a large set of original Mylar drawings was temporarily left by the loading dock at a large shipyard in the Pacific Northwest. The drawings were mistaken for trash and carted away. The drawings were recovered three days later at the local garbage dump.

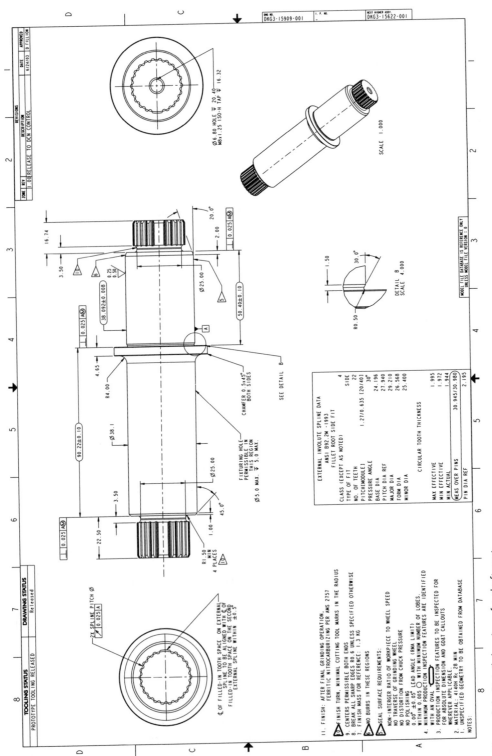

Figure 8-4 Detail drawing of a shaft part

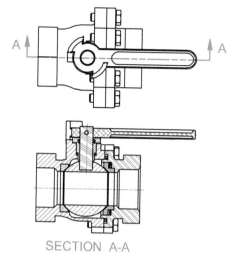

SECTION A-A

Figure 8-5   Assembly section view of a ball valve

nuts, bolts, screws, shafts, pins, keys, bearings, spokes, and ribs. The shaft and nuts on the globe valve appearing in both Figures 8-6 and 8-7 are unsectioned, for example.

- Extremely thin parts, like gaskets and sheet metal parts, are shown solid, rather than sectioned.

An exploded view, like the one of the ball valve shown in Figure 8-8 on page 224, is also fairly easy to extract from a parametric assembly model. Since exploded views are easy to visualize, clearly showing the different parts and how they fit together to form an assembly, these views are commonly found in installation manuals and part catalogs.

When creating an assembly drawing, use the minimum number of views necessary to describe the part relationships. Often a single view is sufficient. It is not necessary to show individual parts in detail, since this is accomplished in the detail drawings. Standard parts are shown in the assembly views. Dimensions are not shown in an assembly drawing, unless they pertain directly to the assembly as a whole. Hidden lines are typically not shown in an assembly drawing.

In addition to the assembly view showing all of the parts, the other main elements of an assembly

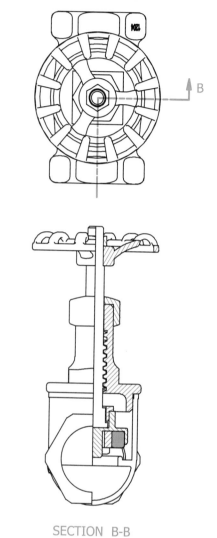

SECTION B-B

Figure 8-6   Half-section view of a globe valve

drawing include a parts list providing information about these parts, and balloons used to identify the parts and to relate them to the parts list. The parts list and balloons are discussed in the following section.

□ BILL OF MATERIALS AND BALLOONS

The bill of materials (BOM), or *parts list*, provides a tabular list of information about the individual parts in the assembly. As shown in

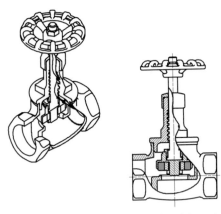

Figure 8-7   Broken-out section view of a globe valve

ITEM	QTY	PROCESS	DESCRIPTION	SIZE	MAT'L	PART/DWG NO
1	1	.	AIR CYLINDER 2 X 7	ORTHO U4SBC	T304SS	0S-001-0015
2	1	.	AIR CYLINDER MOUNT			DKXNP1123
						MBXNXX110C ⚠
3	1	.	DOOR			DKXNP116C
4	1	.	DOOR HINGE			DKXNP118A
5	4	.	HINGE EAR			
6	2	⚠	BOLT, HEX HD - ⌀7/16-14NC X	3	T304SS	03-018-0070
7	1	.	TOGGLE LINK PLATE			MBXLX113B
8	1	.	ALL THREAD	#5/8-18 X 8 1/4	T304SS	03-018-051B
9	2	.	CLEVIS	FOR 2" ORTHO CYL.	T304SS	0S-001-0150
10	1	.	TOGGLE HINGE PLATE			DKXNP119A
11	2	90"MITER45'	PIPE	2 SCH 40 X 24 5/32	T304SS	01-005-0020
12	1	45"MITER45'	PIPE (FOR BOTH DOORS)	2 SCH 40 X 72 3/8	T304SS	01-005-0020
13	1	.	FLASHING LIGHT			SPXXX9199B
14	1	.	FLASHING LIGHT BOX MOUNT			
15	6	. ⚠	BOLT, HEX HD - ⌀7/16-14NC X	2	T304SS	03-018-0066
16	1	.	DISCHARGE GUARD			VMXNX801C
17	1	.	BEARING DRIP PAN			VMXNX318C
18	1	.	FRONT SEAL DETAIL ASSEMBLY			VMXNX400C
19	8	. ⚠	NUT, NYLOCK - 7/16-14NC			03-018-0375
20	2	. ⚠	PILLOW BLOCK BEARING	SEALMASTER RPB-212-C2CR		03-001-0084B
21	1	90"/COPE	PIPE	2 SCH 40 X 22 3/8	T304SS	01-005-0020

Figure 8-9   Parts list for a specific product (Courtesy of Cozzini Inc.)

Figure 8-9, a parts list includes an item number, a unique part identification number, a brief verbal description of the part, the quantity found in the assembly, the material, and possibly other information like weight or stock size. Note that standard parts are also included in the parts list.

The parts list generally appears on the right side of the assembly drawing. Parts are listed in order of importance and/or size. If the column headings appear at the top of the parts list, then the most important parts appear at the top of the list. On some drawings the column headings appear at the bottom, in which case the most important parts appear at the bottom of the list. This practice allows new, presumably less important parts to be added without affecting the item numbering.

Balloons, like those shown in Figure 8-10 on page 225, are used to identify parts and relate them to the parts list. Each balloon consists of a circled number and a leader line. The leader line is drawn to a specific part. The circled number is the item (or find) number for the part, and is used to locate the part in the parts list. The balloons should be neatly organized, preferably in horizontal or vertical rows. Balloon leaders should not cross, and adjacent leaders should be parallel.

Once an assembly model is available, a parts list can be automatically generated and customized using parametric modeling software. A balloon tool is also generally available.

☐ SHEET SIZES

Standard drawing sheet sizes for both metric and English units are shown in Table 8-1 on page 225. These and other engineering graphics standards and conventions are maintained in publications by the American Society of Mechanical Engineers (ASME), and are approved by the American National Standards Institute (ANSI).

☐ TITLE BLOCKS

Every drawing contains a *title block* similar to the one shown in Figure 8-11 on page 225. The purpose of the title block is to organize the

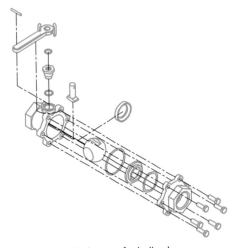

Figure 8-8   Exploded view of a ball valve

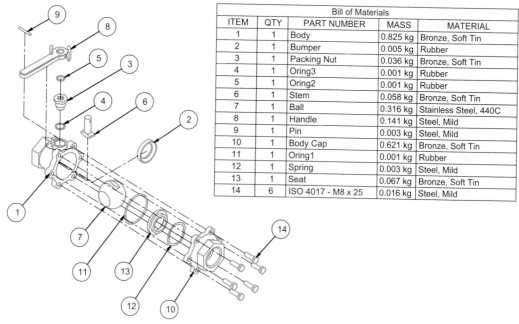

Figure 8-10   BOM, balloons used to identify parts

Bill of Materials				
ITEM	QTY	PART NUMBER	MASS	MATERIAL
1	1	Body	0.825 kg	Bronze, Soft Tin
2	1	Bumper	0.005 kg	Rubber
3	1	Packing Nut	0.036 kg	Bronze, Soft Tin
4	1	Oring3	0.001 kg	Rubber
5	1	Oring2	0.001 kg	Rubber
6	1	Stem	0.058 kg	Bronze, Soft Tin
7	1	Ball	0.316 kg	Stainless Steel, 440C
8	1	Handle	0.141 kg	Steel, Mild
9	1	Pin	0.003 kg	Steel, Mild
10	1	Body Cap	0.621 kg	Bronze, Soft Tin
11	1	Oring1	0.001 kg	Rubber
12	1	Spring	0.003 kg	Steel, Mild
13	1	Seat	0.067 kg	Bronze, Soft Tin
14	6	ISO 4017 - M8 x 25	0.016 kg	Steel, Mild

information needed to identify the drawing. Additional information not given directly on the drawing is also included in the title block. Standard title block information includes the following:

- Company name and address
- Drawing title
- Drawing number
- Names, dates, and release signatures of designer, checker, and supervisor
- Revision number

- Sheet size
- Sheet number
- Scale of drawing
- Other information: project name, client name, material, general tolerance information, heat treatment, surface finish, hardness, estimate weight

The company logo is also frequently found on the title block. On most drawings the title block is

Table 8-1   Standard sheet sizes

Metric (mm)		English (inches)	
A4	210 × 297	A	8.5 × 11.0
A3	297 × 420	B	11.0 × 17.0
A2	420 × 594	C	17.0 × 22.0
A1	594 × 841	D	22.0 × 34.0
A0	841 × 1189	E	34.0 × 44.0
ASME Y14.1M-1995		ASME Y14.1-1995	

Figure 8-11   Title block (Courtesy of Jensen Maritime Consultants, Inc.)

located in the lower right-hand corner of the drawing.

ANSI standard title blocks are available for both metric and English units in various sheet sizes. Companies tend to use their own standard title blocks. These standard title blocks can be made available from a CAD library or as a template, and are easily inserted into a CAD file. In many CAD programs the title block text fields are already completed or set up for easy modification. In parametric solid modeling software, the title block text fields are often associatively linked to the CAD database.

Drawings often contain more information than can be comfortably included on a single sheet. In this case a continuation title block can be used on all sheets after the first sheet. The continuation title block contains less information than the title block on the main sheet. Figure 8-12 shows an example of a continuation title block.

☐ BORDERS AND ZONES

All drawings include a rectangular border. Some drawings also include *zones*. Zones

are regular ruled intervals along the edges of the drawing border that assist in finding information on a complicated drawing. These zones are similar to the zones found on highway road maps, with numerals used horizontally and letters vertically. Figure 8-13 shows a portion of a drawing where a detail view is labeled according to the zone in which it is located.

☐ REVISION BLOCKS

Changes are often made to engineering drawings to account for design changes, customer requests, errors, etc. These changes, or *revisions*, are kept track of in a *revision block* typically located on the upper right-hand corner of the drawing. Figure 8-14 on page 227 shows an example of a revision block. At a minimum, each modification recorded in the revision block includes the date, the name of the person responsible, and a brief description of the change. A revision number, normally placed in a circle or triangle, is associated with each revision and placed both on the revision block and in the area on the drawing sheet where the revision has been made. In a zoned drawing, the zone of the affected area will be listed in the revision block.

☐ DRAWING SCALE

CAD models, whether 2D and 3D, are normally created full-size, without concern for how large

Figure 8-12 Continuation title block (Courtesy of Jensen Maritime Consultants, Inc.)

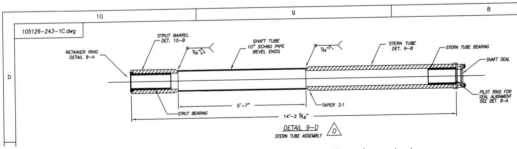

Figure 8-13 Drawing detail labeled by zone (Courtesy of Jensen Maritime Consultants, Inc.)

SYM	ZONE	DESCRIPTION	BY	DATE
		REVISIONS		
PO	ALL	INITIAL ISSUE	YH	4/20 06
P1		1) OVERALL SHAFT LENGTH ADJUSTED PER UNDER WATER SHIPCHECK 4-26-06. 2) RUDDER AND EXISTING STRUT BEARING LOCATED. 3) NEW PROPELLER AND NOZZLE LOCATED TO CLEAR EXISTING RUDDER. 4) TRIMMING OFF 9-½" OF AFT EDGE OF SKEG PLATE SHOWN, TO SUIT THE LONGER CP PROPELLER HUB.	YH	4/29 08
P2		1) NUMBER AND LOCATION OF SHAFT BEARING REVISED. 2) ONE PILLOW BEARING REMOVED. 3) STERN TUBE AND BULKHEAD STUFFING BOXES REPLACED WITH SHAFT SEALS. 4) INDIVIDUAL LENGTHS OF TAIL SHAFT AND LINE SHAFT REVISED (OVERALL LENGTH REMAINS UNCHANGED) 5) SHAFT LINE MATERIALS CHANGED TO CARBON STEEL GR. 4 WITH BRONZE LINERS IN WAY OF STRUT AND STERN TUBE/SHAFT SEAL. 6) SHAFT DIAMETER AND STERN TUBE/STRUT DIAMETERS INCREASED. 7) RESERVATION NOTE #3 REMOVED.	YH	5/3 06
-		1) NEW ENGINE BLOCK RECEIVED FROM CUMMINS INSERTED IN DRAWING. 2) CRAFT MODEL S1 BCH 508 FL PILLOW BLOCK SHOWN. 3) SHAFT LINE ELEVATION (HEIGHT ABOVE BASE LINE) CORRECTED. 4) RESERVATION NOTE #2 REMOVED. GEN. NOTE #2 & #3 REVISED. 5) PRELIMINARY STATUS REMOVED.	YH	5/12 06
A		REVISED PROPELLER HUB FROM HUNDESTED INSERTED.	YH	5/19 06
B		1. LINE SHAFT BEARING RELOCATED PER HUNDESTED SUGGESTION. 2. PROPELLER NOZZLE TILTED 2 DEGREES.	YH	5/31 06
C		1. LINE SHAFT BEARING RELOCATED TO FRAME #39. 2. TAIL SHAFT LENGTH INCREASED, LINE SHAFT LENGTH DECREASED. 3. STERN TUBE AND STRUT DETAILS ADDED (SHT #2).	YH	6/12 06

Figure 8-14 Revision block (Courtesy of Jensen Maritime Consultants, Inc.)

```
NOTES:
UNSPECIFIED TOLERANCES:
DECIMALS:  .0000  ± .0005
           .000   ± .005
           .00    ± .01
FRACTIONS:  ±1/32
ANGLES    ±1°
SQUARENESS AND PARALLELISM
±.015 INCHES PER FOOT
CONCENTRICITY: T.I.R EQUALS
TOLERANCE ON DIAMETER
DIMENSIONING SYSTEM:
ANSI/ASME Y14.5M
```

Figure 8-15 General tolerance note (Courtesy of Cozzini Inc.)

or small the actual object may be. In this way the CAD data is based on the true size of the object. Added dimensions will be true size, query commands will report true size data, and the CAD database will be suitable for export to downstream analysis and manufacturing applications.

Once the full-size model is complete, drawings are prepared and scaled to fit the designated drawing sheet size. Common scales like those available on the engineering scales discussed in Chapter 2 should be used. Table 8-2 provides examples of some common scales used on engineering drawings.

The drawing scale is normally indicated on the title block. In the event that different views on the drawing sheet are drawn at different scales, then the scale of the view is indicated on the view label.

□ TOLERANCE NOTES

Many companies include a general tolerance note similar to the one shown in Figure 8-15 on their detail drawings. These tolerance notes refer to any dimensions that have not been specifically toleranced. Tolerance notes are typically found in the lower right-hand corner of the drawing.

□ STANDARD PARTS

Standard parts may either be purchased from outside or produced within the company. Typical standard parts include threaded fasteners, bushings, bearings, pins, keys, pumps, valves, etc. While standard parts do not require a detail drawing, they are shown on the assembly drawing and included in the parts list. Most 3D parametric modeling programs have a built-in standard parts library, from which standard part models can be directly inserted into an assembly model. Figure 7-37 on page 203 in the previous chapter shows a screen capture of a CAD library interface within a parametric modeling program.

Table 8-2 Common drawing scales

1:1	$1/8'' = 1'\text{-}0''$
1:2	$1/4'' = 1'\text{-}0''$
1:4	$3/8'' = 1'\text{-}0''$
1:8	$1/2'' = 1'\text{-}0''$
1:10	$1'' = 1'\text{-}0''$
1:20	$1'' = 10'$
1:30	$1'' = 20'$
1:40	$¼'' = 1''$
1:50	$½'' = 1''$
1:100	$1'' = 1''$

☐ WORKING DRAWING CREATION USING PARAMETRIC MODELING SOFTWARE

In the remaining pages of this chapter various techniques to develop working drawings are described. These parametric modeling techniques include the creation of part drawings, assembly models, sectioned assembly and exploded views, as well as adding a parts list and balloons to an exploded view.

Extracting a detail drawing from a parametric part model

The steps required to generate a detail drawing of a part using parametric modeling software are described. See Figure 8-16.

S T E P 1. Insert a base view of the part model.

S T E P 2. Using the base view, project the other views.

S T E P 3. Add centerlines.

S T E P 4. Import parametric dimensions from part model. Note that these dimensions will probably need to be repositioned. In addition, at least some of these imported parametric dimensions will not be suitable for documentation purposes, and will need to be replaced.

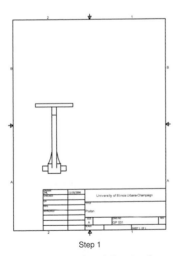

Step 1

Figure 8-16 Extracting a detail drawing from a parametric model

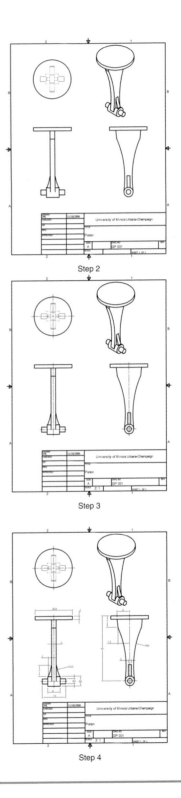

Step 2

Step 3

Step 4

Using existing part models to create an assembly model

The steps required to create a parametric assembly model are described, assuming that all of the part models have already been created. See Figure 8-17.

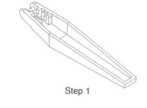

Step 1

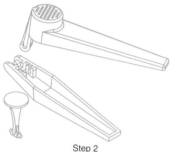

Step 2

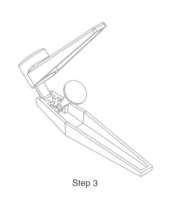

Step 3

Step 4

Figure 8-17 Creating an assembly model

STEP 1. Import the base part into the assembly environment.

STEP 2. Import the other component files into the assembly.

STEP 3. Using assembly constraints like mate and insert, correctly position the components with respect to the base component and to one another.

STEP 4. Properly constrained, moving parts will behave as they would in reality. The garlic press assembly model can be opened and closed, simulating the behavior of the actual device.

Extracting a sectioned assembly drawing

This section shows the steps needed to create a sectioned assembly drawing from a parametric assembly model. See Figure 8-18 on page 230.

STEP 1. Import a base view of an assembly into the drawing environment.

STEP 2. Using a sectioning tool, create the section view(s). Notice that the angle of the section lining (ANSI 31) differs for the different parts through which the cutting plane passes. Note also that some conventions (e.g., do not apply section lining to cut thin features) are not automatically adhered to.

STEP 3. Modify the hatch pattern to adhere to convention. Notice also that the holes in the top of the cylindrical handle part have been suppressed for clarity at this stage.

Creating an exploded view

This section shows the steps required to develop an exploded view of a parametric assembly model. See Figure 8-19 on page 231.

STEP 1. Import an assembly model into the exploded view creation environment.

STEP 2. Translate and/or rotate the parts to simulate the product's disassembly.

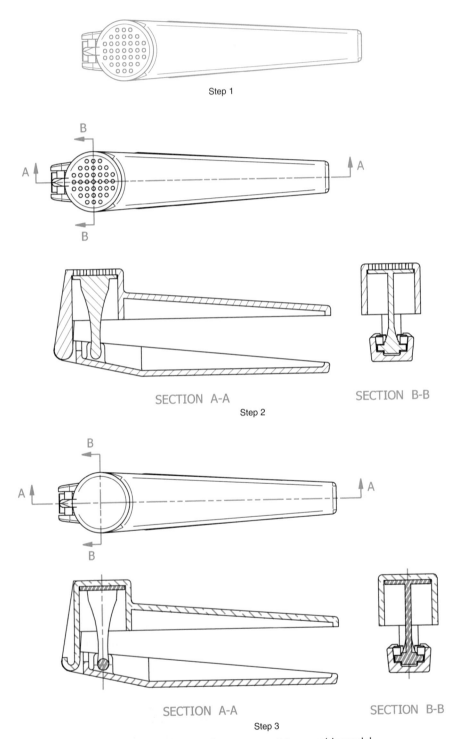

Step 1

B

A ↕   ↕ A

B

SECTION A-A   SECTION B-B

Step 2

B

A ↕   ↕ A

B

SECTION A-A   SECTION B-B

Step 3

Figure 8-18 Extracting a sectioned assembly drawing from a parametric assembly model

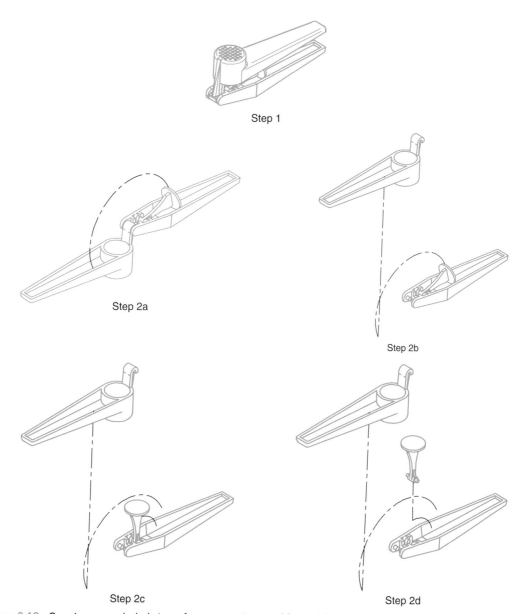

Step 1

Step 2a

Step 2b

Step 2c

Step 2d

Figure 8-19   Creating an exploded view of a parametric assembly model

Creating an exploded view drawing with parts list and balloons

The process of generating an exploded view drawing, parts list, and balloons is described. See Figure 8-20 on page 232.

STEP 1. Inserted an exploded view into the drawing environment.
STEP 2. Insert a parts list that is linked to the exploded view.
STEP 3. Customize the parts list.
STEP 4. Add balloons.

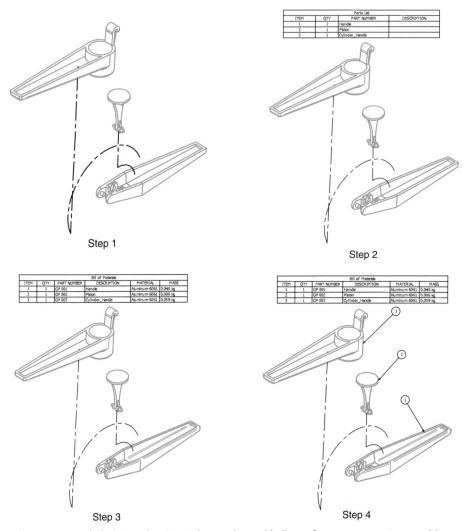

Step 1

Step 2

Step 3

Step 4

Figure 8-20 Creating an exploded view drawing with parts list and balloons from a parametric assembly model

☐ QUESTIONS

1. Given a dimensioned isometric view (P7-1 through P7-24), create a dimensioned detail drawing of the part.

2. Given dimensioned assembly views of an artifact (P8-1 through P8-5), create a complete set of working drawings.

   a. Create a solid model of each part.

   b. Create dimensioned detail drawings for each of the individual parts, including auxiliary views and section views when appropriate.

   c. Create an exploded pictorial assembly drawing, including balloons and a parts list.

   d. Where appropriate, create a sectioned assembly drawing.

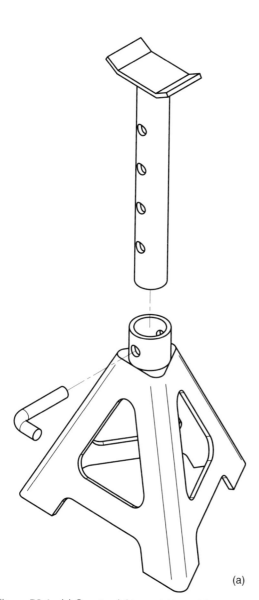
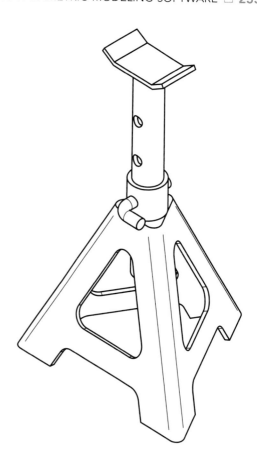

Figure P8-1   (a) Car stand (b) arm (c) base (d) pin

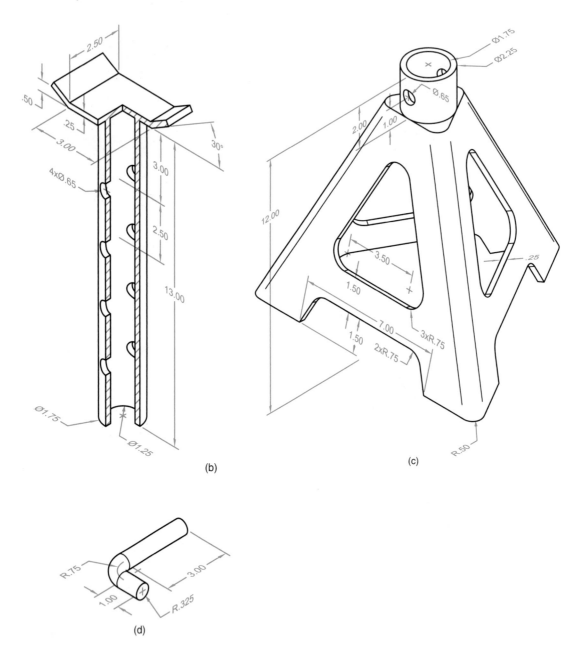

(b)

(c)

(d)

Figure P8-1 (Continued)

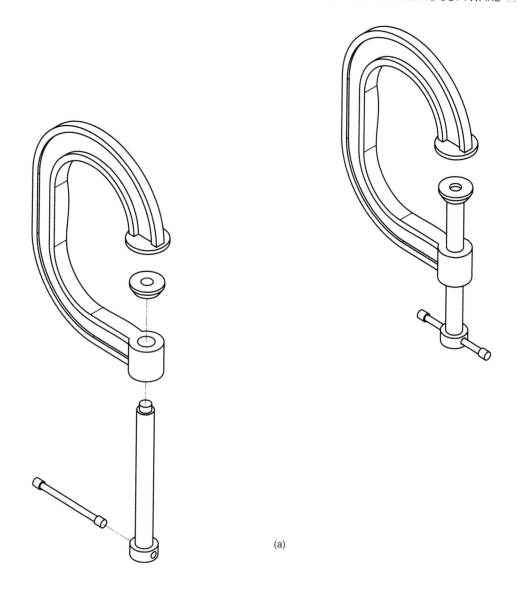

(a)

Figure P8-2　(a) C-clamp (b) base (c) pin (d) screw (e) screw head

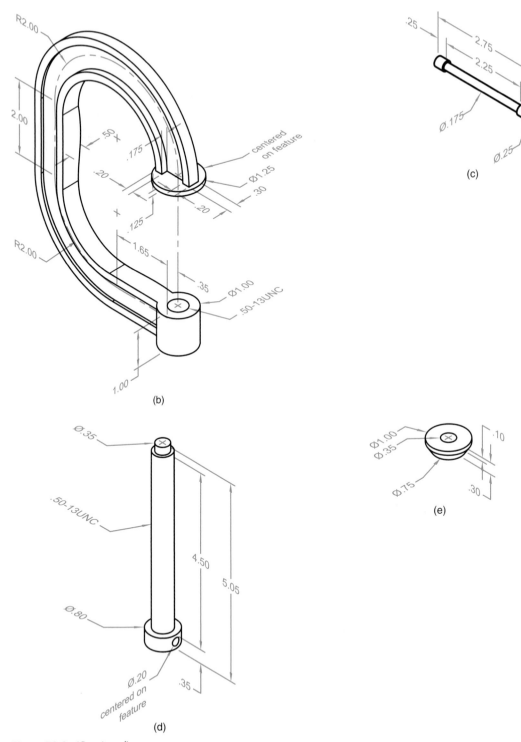

Figure P8-2 (Continued)

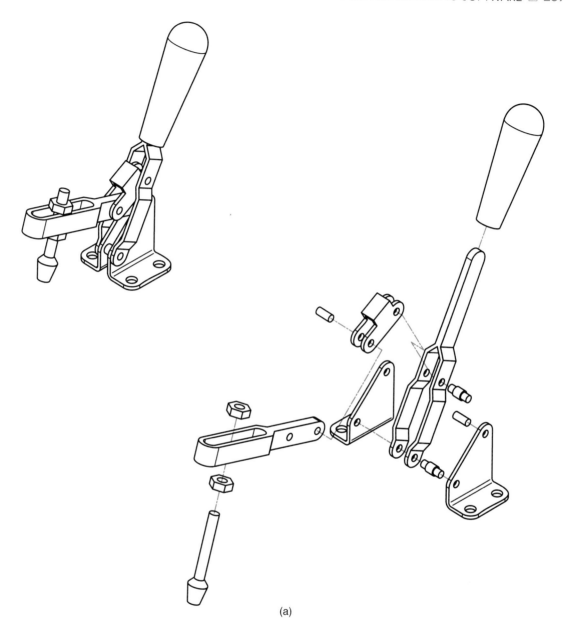

(a)

Figure P8-3   (a) Manual clamp (b) arm (c) grip (d) handle (e) left support (f) lever (g) nut (h) pin A (i) pin B (j) stop

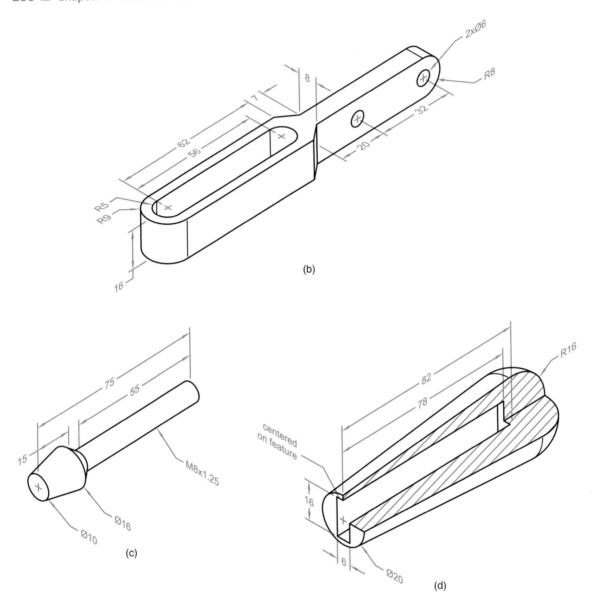

Figure P8-3 (Continued)

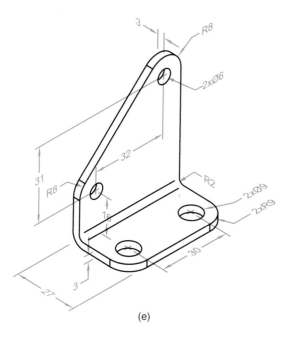

(e)

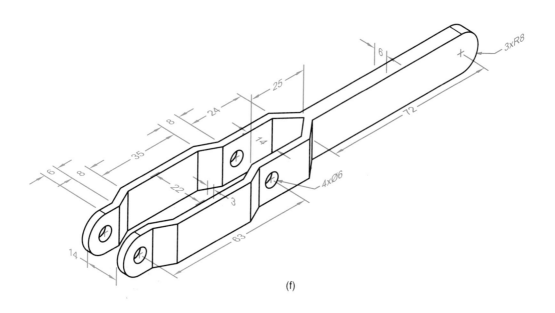

(f)

Figure P8-3 (Continued)

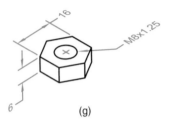

(g)

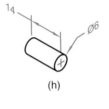

(h)

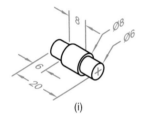

(i)

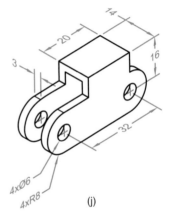

(j)

Figure P8-3 (Continued)

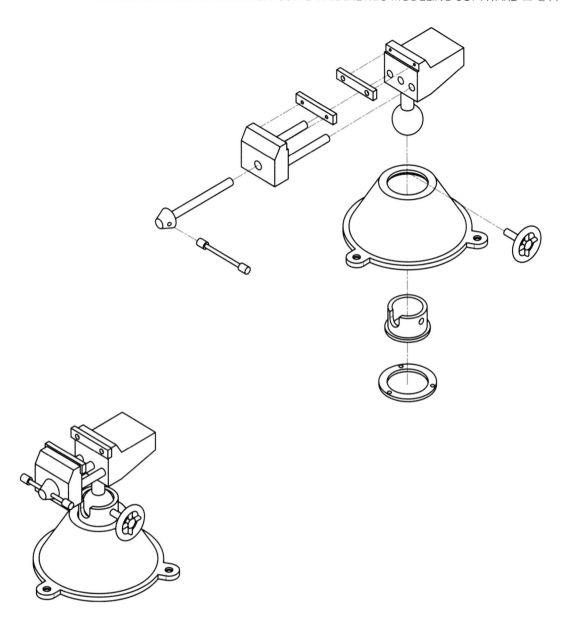

Figure P8-4   (a) Vise

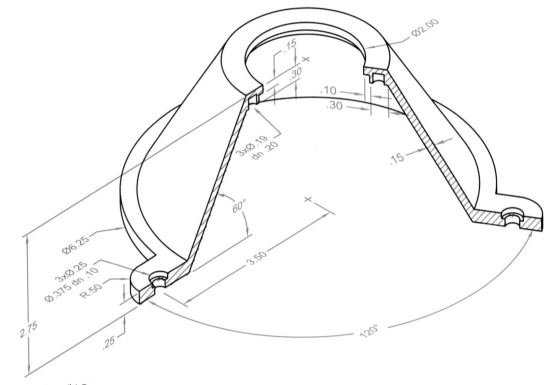

Figure P8-4 (b) Base

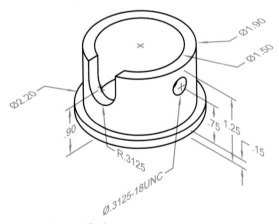

Figure P8-4 (c) Casing

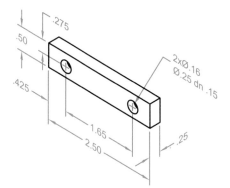

Figure P8-4 (d) Grip

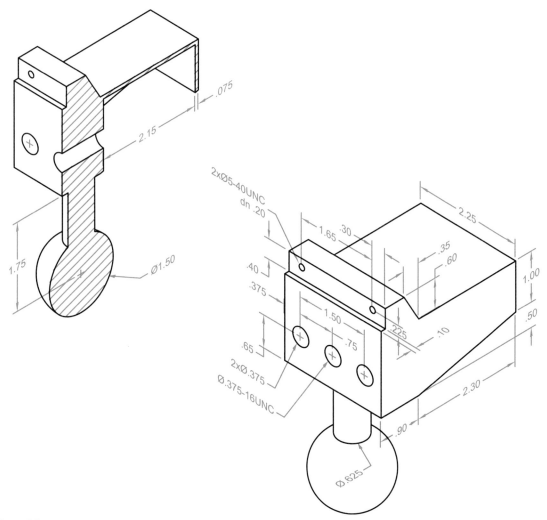

Figure P8-4 (e) Head

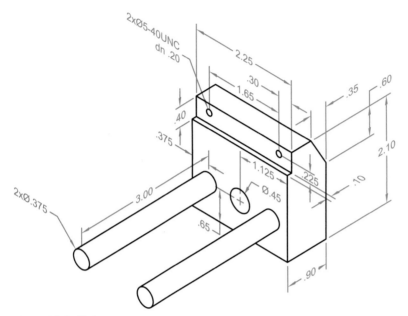

Figure P8-4 (f) Jaw

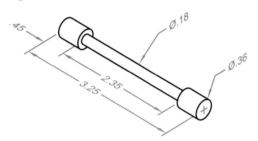

Figure P8-4 (g) Pin

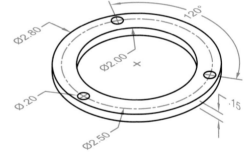

Figure P8-4 (h) Plate

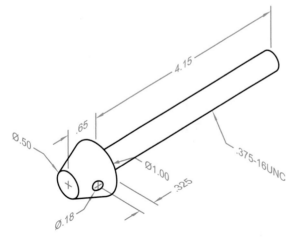

Figure P8-4 (i) Shaft

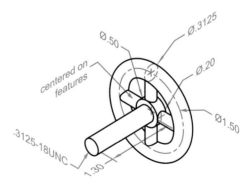

Figure P8-4 (j) Wheel

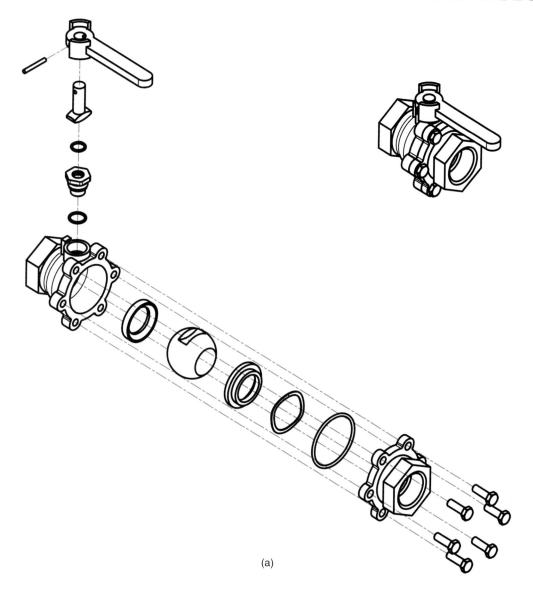

(a)

Figure P8-5   (a) Ball valve (b) ball (c) body cap (d) body (e) bumper (f) handle (g) O-ring 1 (h) O-ring 2 (i) O-ring 3 (j) packing nut (k) pin (l) seat (m) spring (n) stem

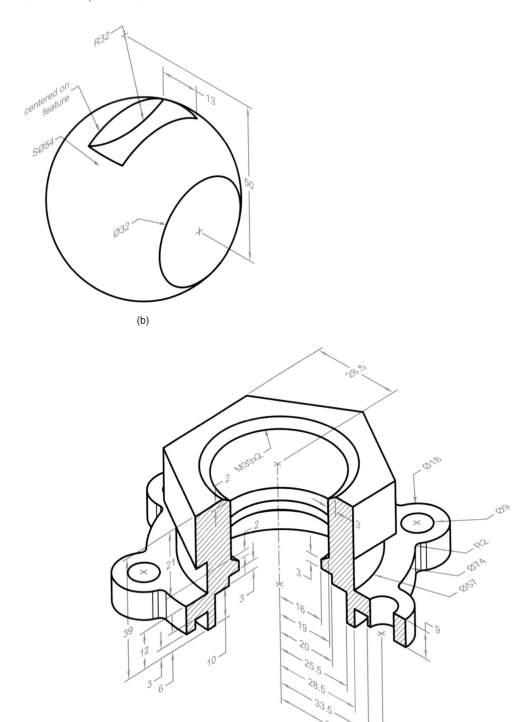

(b)

(c)

Figure P8-5   (Continued)

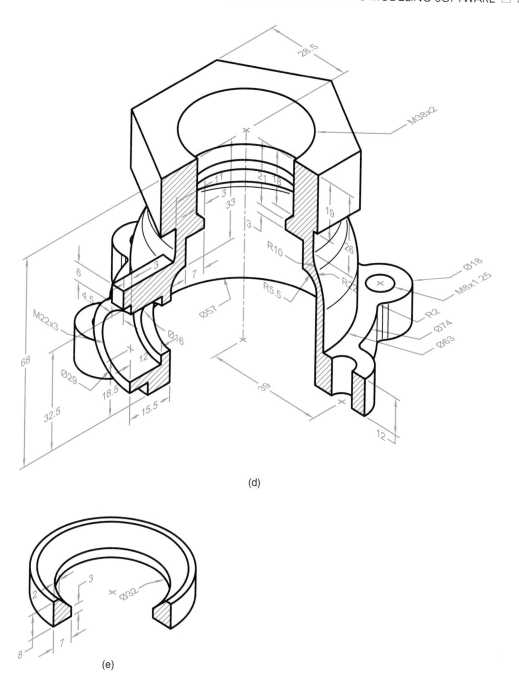

(d)

(e)

Figure P8-5 (Continued)

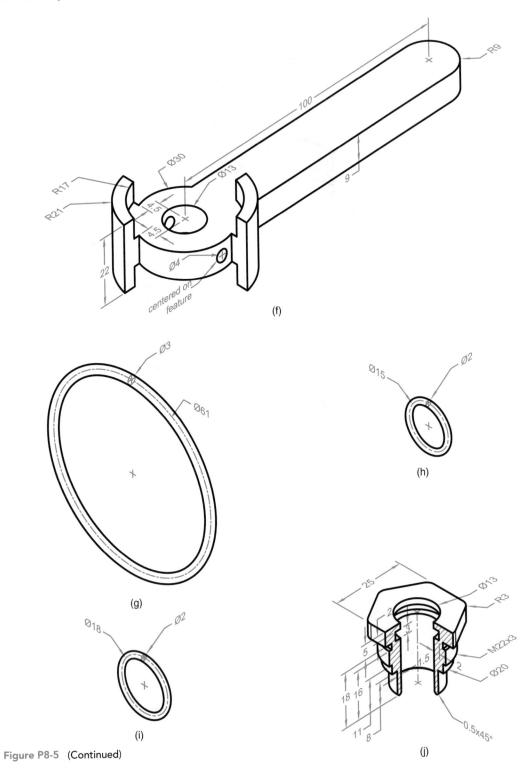

Figure P8-5 (Continued)

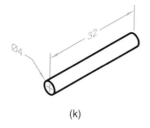

(k)

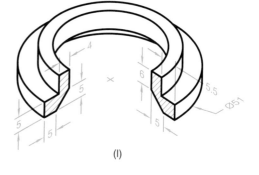

(l)

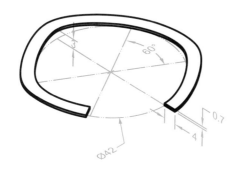

uncompressed height is 7
compressed height is 3

(m)

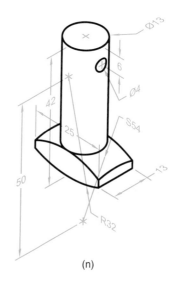

(n)

Figure P8-5 (Continued)

9 PRODUCT DISSECTION, REVERSE ENGINEERING, AND REDESIGN

□ INTRODUCTION

Chapter 1 began with the observation that design is the central activity of the engineering profession. The chapter then attempted to describe the nature of engineering design: what it is, how it is done, etc. Engineering design is truly about finding solutions to *real-world* problems. Good design calls for verbal (common sense, reasoning), graphical (sketching, CAD modeling), social (communication, collaboration, teamwork), as well as analytical (mathematical, scientific) skills focused on finding innovative solutions to open-ended problems.

Engineering design is difficult to master. There is no *cook-book approach* to design. Fortunately, there is another design truism, "all design is redesign," that provides another avenue for learning design other than "just do it." Young engineers are frequently surprised (and perhaps disappointed) to find how rare original design is. Very few design projects start with a blank sheet of paper. Most design efforts begin with an existing design or product that serves as the basis for the new project. In recent years, given today's global market, this reality has become more evident. In order to remain competitive, modern manufacturing companies must roll out new products (e.g., digital cameras, computers, cell phones) with astonishing frequency.

The subject of this chapter is product dissection and, more broadly, reverse engineering. These techniques are based on the notion that by taking apart and studying mechanical or electrical devices, we can better understand not only the specific product, but also the design process that produced it. After all, one of the biggest obstacles to learning design is that we are forced to begin with a blank sheet of paper. In starting with an existing product that is disassembled, analyzed, sketched, modeled, and finally improved upon, hands-on experience is added to our understanding of design.

□ PRODUCT DISSECTION VERSUS REVERSE ENGINEERING

Product dissection is an approach to learning about engineering concepts and design principles by exploring engineered products. In a product dissection project, the relationship between the form and function of a device is investigated. Also called mechanical dissection, product dissection allows us to see how others have successfully solved a particular design problem.

In a typical dissection exercise, student teams disassemble, study, and then reassemble commercial products like bicycles, power tools, and coffee makers. Frequently cited benefits of this hands-on activity include 1) improved mechanical aptitude, 2) basic knowledge of manufacturing processes, materials and material selection, and product decomposition hierarchy, 3) exposure to the design process, 4) appreciation for how ergonomics influences product design, and 5) awareness of how customer needs are translated into product functions that are then converted into commercial products.

While related to product dissection, **_reverse engineering_** is a broader, more systematic approach used to analyze the design of existing devices or systems, often with the aim of duplicating or enhancing the device. Reverse engineering[1] techniques are commonly used in industry, either to benchmark a competitor's product or as a first step in a product's redesign. Reverse engineering also refers to the process of creating a 3D CAD model of a part when the part drawing is not available. The part is measured and then modeled in the CAD program. A coordinate measuring machine (CMM) or 3D digitizer, like the one shown in Figure 9-1, may also be used to reverse engineer the part.

World-class companies employ reverse engineering in order to gain competitive advantage. In particular, reverse engineering is used 1) to understand a product's design, 2) to promote product improvement, and 3) for competitive benchmarking. Reverse engineering techniques offer a systematic approach to fully understanding and representing a product's design, either

Figure 9-1 3D digitizer (Reproduced by permission of Immersion Corporation, Copyright © 2007 Immersion Corporation. All rights reserved)

---

[1] When used in reference to software, the term "reverse engineering" refers to reversing a program's machine code back to the original source code that it was written in, using program language statements. Software reverse engineering is used when the original source code is not available.

one's own or that of a competitor. It also provides a means to explore new avenues for the continuous improvement, evolution, and redesign of a product or product family. Finally, reverse engineering is a benchmarking technique that allows a comprehensive understanding of competitor products. This knowledge can then be leveraged to identify *best-in-class* approaches to common problems, which are then used to develop better products.

□ PRODUCT SUITABILITY

The following guidelines may be used when selecting a product or mechanism suitable for disassembly:

- Assemblies with from 10 to 15 parts are ideal, but a range from 5 to 30 parts is workable
- Try to select products with moving parts, since CAD software animation and motion analysis capabilities can then be used
- Look for products that can easily be disassembled and reassembled, either by hand or with simple tools; avoid products with welded housings and glued joints
- Try to select an inexpensive product; if it must be broken to disassemble, a second one can be purchased
- The device should still be in working order, so that it is possible to see how the external functionality is achieved internally
- The artifact should be easily portable; it should not be large or heavy, nor should it be too small
- The product should not require complicated assembly instructions
- You should *want* to reverse engineer the product
- Ideally, the internal functioning of the product should not be readily apparent.
- The product preferably should offer the potential for improvement.

Figure 9-2 shows several commercial products that meet these requirements.

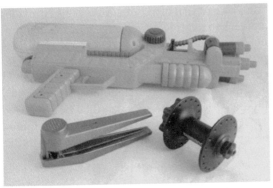

Figure 9-2 Commercial products used for product dissection projects: bicycle front hub, kitchen item, toy

Figure 9-3 Craftsman Professional locking pliers (Courtesy of Craftsman Tools, Inc.)

☐ PRODUCT DISSECTION PROCEDURE

In addition to accumulating knowledge of mechanisms and product design, product dissection experiences also offer many opportunities to develop graphical and visualization skills. Product dissection activities can be expanded to include freehand sketching, instrument drawing, and CAD modeling. Assuming that a parametric assembly model of the product has been created, the virtual prototype can be documented (working drawings, rendered views, physical prototypes), and analyzed (weight estimate, tolerancing, motion analysis, stress analysis). In an effort to continuously improve their products, manufacturing companies routinely employ reverse engineering practices. In an academic setting, a product improvement component can likewise be incorporated into the dissection experience. Finally, as with other design projects, it is important that the dissection team communicate its findings. An outline of a product dissection project incorporating these features is shown below:

- Pre-dissection analysis
- Dissection
- Product documentation
- Product analysis
- Product improvement
- Product reassembly
- Communication

In the course of the subsequent discussion, a pair of locking pliers made by Craftsman Tools (see Figure 9-3) will be used to illustrate the dissection process. Locking pliers are used by trade professionals, as well as homeowners, for clamping, tightening, twisting, and turning.

☐ PRE-DISSECTION ANALYSIS

Prior to taking the product apart, it is worthwhile to evaluate the product's functionality and market segment. With regard to functionality, the device should be observed in operation. What does the device do? Make some predictions regarding how the device works. How many parts does the product contain? What sort of mechanisms allow the device to work as it does? What scientific principles were used in the design of the mechanism?

Locking pliers like the Craftsman Professional 7-inch locking straight jaw version are popular for welding, metalwork, and other trades. They are also found in home tool boxes. This multipurpose hand tool functions as an adjustable wrench, but can also be used as a clamp or vise capable of holding a workpiece in a fixed position. The primary mechanism employed in locking pliers is a *four bar linkage*, although the principle of *leverage* is also used. The locking pliers appear to contain about ten different parts.

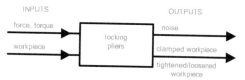

Figure 9-4 Black-box representation of locking pliers' functionality

Next, create a black-box diagram that represents the overall intended function of the product. Inputs and outputs can either be categorized as material, energy, or signals (information). A black-box representation showing the functional structure of the locking pliers is shown in Figure 9-4.

A list of customer needs should be identified to establish the market segment. Who are the target customers? Who are the principal competitors? Identify some features of competing products. What are the advantages and disadvantages of this product versus that of the competition?

A list of locking pliers' customer needs might include the following:

- Tough, durable
- Reliable
- Easy to adjust
- Easy to use
- Can be locked in any position
- Easy to release locking mechanism
- Comfortable grip
- Affordable

While the primary customers for locking pliers are welders and other metal workers, the basic design can and has been modified to expand the target user base to include other crafts. The Craftsman locking pliers appear to have one principal competitor, and several smaller competitors who typically produce less expensive versions of the locking pliers' concept. While most of the designs include a trigger component that is used to release the locking mechanism, the Craftsman tool has a patented design that permits an easier release without the need for the trigger component. The Craftsman tool also includes a non-slip, cushion grip for increased comfort and reliability.

□ DISSECTION

In dissecting the product, the principal goal is to understand how the product works and is assembled. Another goal is to identify potential improvements that can be made to the product.

As was mentioned earlier, potential product dissection candidates should be easy to take apart, either by hand or with simple tools. Useful tools include screwdrivers, wrenches, and pliers. In difficult situations electric drills, shear cutters, a hammer, or a hacksaw may need to be used. The Craftsman locking pliers proved difficult to disassemble, owing to its rugged design. Steel rivets are used to connect the locking pliers' components, which had to be removed.

Good lighting is certainly important when dissecting products. A workbench equipped with a vise or clamps is also helpful. A camera is useful in order to document the dissection process, especially when used to show the order and orientation of parts as assembled (i.e., exploded views). Figures 9-5 through 9-13 starting on page 254 show photographs of the disassembled components of the locking pliers' dissection process.

As the product is disassembled, it is a good idea to keep a written record of the disassembly sequence. This product disassembly plan can later be used to help create an exploded view and an assembly animation file. If at all possible, try to avoid breaking parts. If breakage is unavoidable and the product is inexpensive, a second product may be purchased so that a working version is available.

Craftsman locking pliers disassembly steps

The following steps provide a disassembly sequence for the locking pliers shown in Figure 9-3:

1. Unscrew the adjusting screw from the fixed handle. Note that as the adjusting screw is removed, the jaws of the locking pliers open (Figure 9-5).

2. Free the end of the toggle link from the fixed handle channel by pushing down on the toggle link hump feature (thus closing the jaws), and then rotating the link end out of the channel (Figure 9-6).

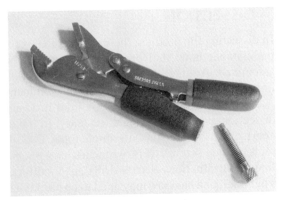

Figure 9-5 Adjusting screw removed

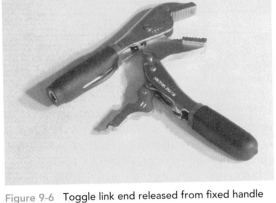

Figure 9-6 Toggle link end released from fixed handle channel

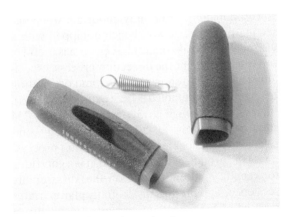

Figure 9-7 Cushion grips and spring

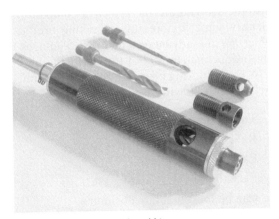

Figure 9-8 Rivet removal tool kit

3. Using a small screwdriver or other tool, remove the spring that spans between fixed handle channel and the jaw. Note that the spring serves to resist the closing of the opposable jaws.

4. Using a flat screw driver or other tool to stretch and loosen them, slide the cushion grips from both the fixed and movable handles. Figure 9-7 shows the cushion grips and the spring.

5. After clamping the fixed handle in a bench vise, use an electric drill, a rivet removal tool (see Figure 9-8), and regular drill bitts to remove the head of the rivet connecting the fixed handle to the jaw (Figure 9-9). Once the rivet head has been removed, use a hole punch and hammer to completely remove the rivet. Figure 9-10 shows the fixed handle subassembly components.

6. Use the process described in step 5 above to remove the rivet connecting the jaw to the movable handle subassembly.

7. Repeat the process described in step 5 above to remove the rivet connecting the movable handle to the compound toggle link subassembly. Figure 9-11 shows the movable handle subassembly components.

8. Figure 9-12 shows the compound toggle link subassembly. The flat rivet connecting the compound link to the toggle link should be removed using the same technique described in step 5. In practice however, this proved to be too difficult. It was not, however, difficult to obtain the dimensions of both parts while still assembled (see Figure 9-17 on page 257, which

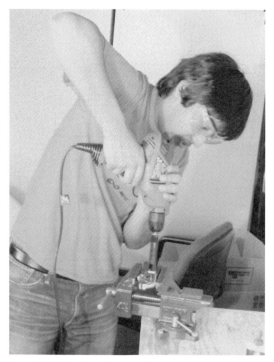

Figure 9-9   Rivet head removal

Figure 9-10   Fixed handle subassembly components

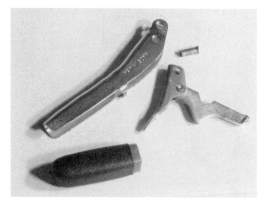

Figure 9-11   Movable handle subassembly components

Figure 9-12   Compound toggle link subassembly

Figure 9-13   Disassembled parts for the locking pliers

includes a tracing of the subassembly profile). Figure 9-13 shows the disassembled locking pliers components.

□ PRODUCT DOCUMENTATION

Product documentation can include a number of elements, among them a product decomposition diagram, freehand sketches of the parts and assembly, and a parametric assembly model of the product. If the product is modeled, then the list of documentation items can be expanded to include working drawings, 3D prints, and rendered views.

A product component decomposition diagram, as seen in Figure 9-14 on page 256, can be used to capture the product's subassembly and

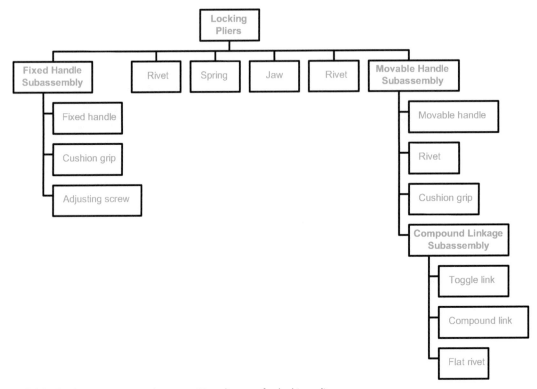

Figure 9-14 Product component decomposition diagram for locking pliers

part relationships. Note that in order to construct the diagram it is first necessary to identify appropriate part names, another useful activity.

Product dissection exercises provide the opportunity to use the sketching techniques discussed in earlier chapters on these open-ended problems. Depending upon the dissected product, any number of sketches may be appropriate, including multiviews, pictorial views (isometric, oblique, perspective), section views (full, offset, half, broken-out, revolved, removed), and auxiliary views. Assembly sketches (e.g., section, broken-out) can also be executed, although exploded views are best done within CAD software. Figure 9-15, for example, shows several section views of a pump head part, made in the course of dissecting a portable water filter product.

While tools needed to execute the part sketches are normally limited to pencil and paper, digital calipers (see Figure 9-16) are also

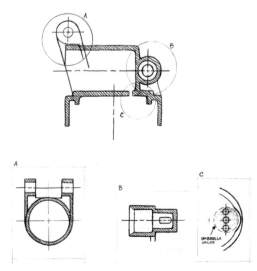

Figure 9-15 Section view sketches of portable water filter pump head (Courtesy of Andrew Block)

Figure 9-16 Digital calipers

necessary if the sketches are to include the dimensions needed to create an accurate CAD model of the product. In this case, the opportunity also exists to put previously acquired dimensioning and even tolerancing skills into practice.

Figure 9-17 displays traced sketches of three different locking pliers' parts to be used for dimensional take-off. Note that a circle template and triangles have been used in the execution of these sketches. Also notice that hole features are used as the origin in two of these part sketches. Compare the sketch of the jaw part with the parametric sketch of the same part appearing in Figure 9-18 on page 258. One of the critical hole features is located at the sketch origin, while the second hole is on the y axis.

Once the parts are modeled, subassemblies and a final product assembly model may also be created, assuming that parametric solid modeling software is being used. In the event that some parts have complex, sculpted surfaces, it may be possible to use a 3D digitizer (see Figure 9-1 on page 251), scanner, or coordinate measuring machine to capture the part's surface features. Figure 9-19 on page 258 shows several CAD models of different locking pliers' parts.

Assuming that the dissected product has been modeled (parts, subassemblies, final product assembly), working drawings can also be derived.

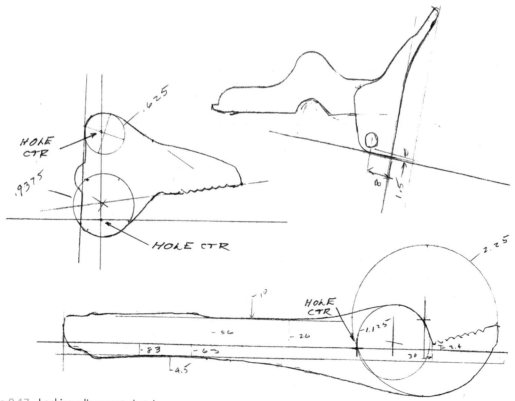

Figure 9-17 Locking pliers part sketches

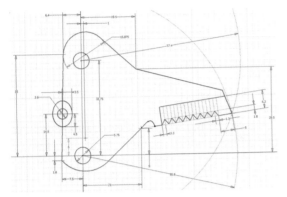

Figure 9-18 Parametric sketch of jaw part

In addition to dimensioned and annotated part drawings (see Figure 9-20), an assembly drawing including assembled views (see Figure 9-21) as well as an exploded view (see Figure 9-22 on page 259) can all be developed. Note that in the process of creating the exploded view most parametric modeling software programs also produce an animation file. This animation file documents the product's assembly, since the disassembly steps are played in reverse order. A parts list or bill of materials (BOM) can also be created in association with the exploded or other assembly views. See Figure 9-23 on page 260.

If a 3D printer is available, additional product documentation can include a physical prototype of individual parts, a critical subassembly, or even the entire product. Figure 9-24 on page 260 shows a 3D print of the assembled locking pliers' parts. Assuming that a rendering module is either included within the CAD system or available separately, a photorealistic rendering of the product may also be created. Figure 9-25 on page 260 shows a rendered view of the locking pliers.

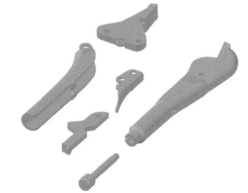

Figure 9-19 CAD models of locking pliers' parts

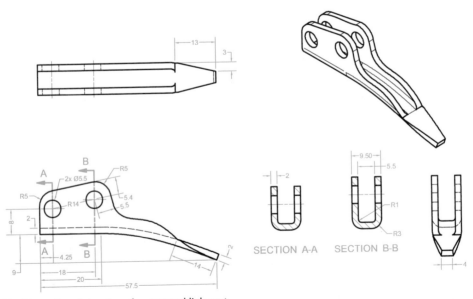

SECTION A-A   SECTION B-B

Figure 9-20 Dimensioned drawing of compound link part

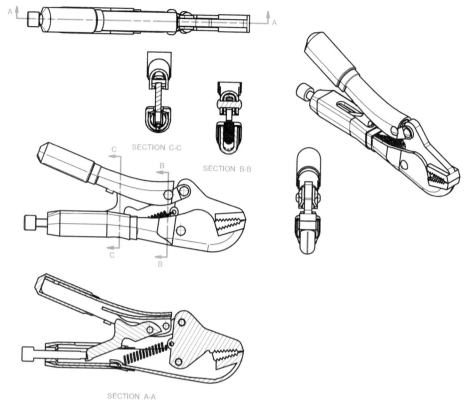

SECTION C-C

SECTION B-B

SECTION A-A

Figure 9-21 Assembly drawing views of locking pliers

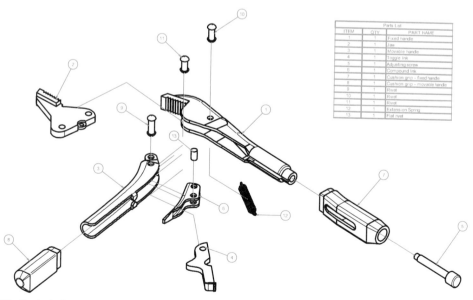

Parts List		
ITEM	QTY	PART NAME
1	1	Fixed handle
2	1	Jaw
3	1	Movable handle
4	1	Toggle link
5	1	Adjusting screw
6	1	Compound link
7	1	Cushion grip - fixed handle
8	1	Cushion grip - movable handle
9	1	Rivet
10	1	Rivet
11	1	Rivet
12	1	Extension Spring
13	1	Flat rivet

Figure 9-22 Exploded view of locking pliers

Parts List		
ITEM	QTY	PART NAME
1	1	Fixed handle
2	1	Jaw
3	1	Movable handle
4	1	Toggle link
5	1	Adjusting screw
6	1	Compound link
7	1	Cushion grip - fixed handle
8	1	Cushion grip - movable handle
9	1	Rivet
10	1	Rivet
11	1	Rivet
12	1	Extension Spring
13	1	Flat rivet

Figure 9-23 BOM for locking pliers

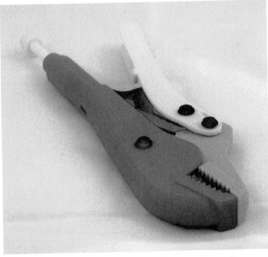

Figure 9-24 3D print of the x and y parts from the locking pliers

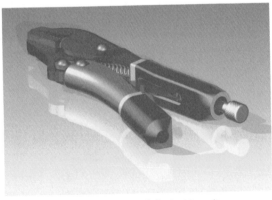

Figure 9-25 Rendered view of the locking pliers

☐ PRODUCT ANALYSIS

Once the dissected product has been documented (using part decomposition diagrams, sketches, models, 3D prints, and rendered views), the device can also be analyzed. Some analysis options include functional decomposition analysis, material analysis and weight estimation, manufacturing process analysis, kinematic analysis, and stress strain analysis.

The goal of functional decomposition analysis is to break down or decompose the primary product function hierarchically into different subfunctions. The product's basic function describes the relationship between the available inputs and the outputs of a product. As seen in Figure 9-4 on page 253 the product function can be described graphically as a black box. Inputs are shown entering on the left, with outputs exiting on the right. Verbally, product functions take the form of an action verb and noun combination. For example, the overall product function of the locking pliers is *clamp workpiece* or *tighten (loosen) workpiece.*

In order to successfully realize the basic product function, a number of secondary functions, or **subfunctions**, must be satisfied. Like the basic product function, these subfunctions can be expressed as action verb + noun combinations. Careful observation of the functioning of the locking pliers, for example, reveals that:

1. **Adjust jaws**–the adjusting screw is first used to set the jaw size (i.e., distance between jaws) to slightly less than that of the workpiece to be clamped.

2. **Open jaws**–the locking pliers are then opened by pushing the movable handle away from the fixed handle.

3. **Clamp workpiece**–after placing the jaws around the workpiece, the jaws are closed by pushing the movable handle towards the fixed handle, effectively clamping the workpiece in place.

4. **Tighten/loosen workpiece**–if required, the workpiece can now be tightened, loosened, twisted, or turned.

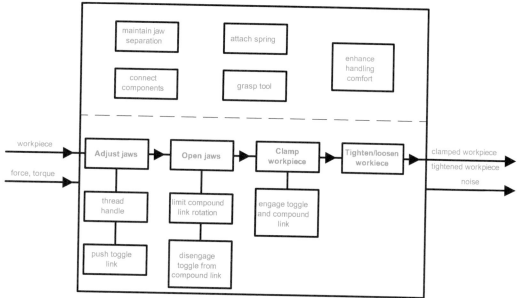

Figure 9-26   Functional decomposition of the locking pliers

5. **Open jaws**–to release the workpiece, the movable handle is once again pushed away from the fixed handle, releasing the workpiece.

In the function structure of the locking pliers shown in Figure 9-26, these **critical-path sub-functions** appear in sequence across the middle of the box, connected by arrows.

In addition to these critical-path subfunctions, a number of other secondary functions must also be addressed. Some of these other secondary functions serve in support of the various critical-path functions. For example, in order to open the locking pliers' jaws, the range of rotation of the compound link must be limited inside the movable handle channel, at which point the toggle link disengages from the compound link, allowing the jaws to open.

Other subfunctions are shown at the top of the box in Figure 9-26. These secondary functions are sometimes referred to as all-time functions, since they must be met at all stages in the operation of the device. In the case of the locking pliers, the principal components, which together form a

four bar linkage, must be permanently connected and yet free to rotate with respect to one another.

Note that there are a number of different ways to graphically represent the functional structure of a product, with Figure 9-26 being one example. The goal of functional decomposition is to gain insight into how a basic product function is realized. By decomposing an overall function into several subfunctions, the design problem becomes more manageable.

Notice also that there tends to be a mapping between subfunctions and components. As an alternative to or in conjunction with a functional decomposition diagram, it is useful to identify the function of the different components. For example, in the locking pliers solid rivets are used to connect the various components, a spring connecting the movable jaw to the fixed handle tends to keep the jaws separated, and handles are provided so that the tool can be grasped.

The main locking pliers' components are *forged*, using high grade, heat-treated alloy steel. The threads of the adjusting screw have been *roll threaded*. The cushion grips are made of rubber,

ITEM	QTY	PART NAME	MATERIAL	MASS (grams)
1	1	Fixed handle	Steel, High Strength Low Alloy	156.06
2	1	Jaw	Steel, High Strength Low Alloy	79.32
3	1	Movable handle	Steel, High Strength Low Alloy	77.70
4	1	Toggle link	Steel, High Strength Low Alloy	27.48
5	1	Adjusting screw	Steel, Mild	20.20
6	1	Compound link	Steel, High Strength Low Alloy	16.70
7	1	Cushion grip - fixed handle	Rubber	9.82
8	1	Cushion grip - movable handle	Rubber	6.42
9	1	Rivet	Steel, High Strength Low Alloy	4.22
10	1	Rivet	Steel, High Strength Low Alloy	3.61
11	1	Rivet	Steel, High Strength Low Alloy	3.27
12	1	Extension Spring	Steel	1.80
13	1	Flat rivet	Steel, High Strength Low Alloy	1.77
			Estimated weight	408.35 grams

Figure 9-27 Weight estimate of the locking pliers

and have been manufactured using *injection molding*.

Using the solid assembly model, it is possible to estimate the virtual product's weight and then compare it with the actual product weight. A weight estimate based upon the reverse-engineered CAD assembly model of the locking pliers appears in Figure 9-27. The estimated weight of 408 grams is reasonably close to the actual weight of 372 grams. Likely reasons for the difference between the estimated and actual weight include modeling and material density inaccuracies.

Assuming that a virtual assembly model has been created and the assembly constraints have been properly defined, animations showing the motion of any moving parts within the product assembly can be displayed. The range of motion of these moving parts can also be determined. Figure 9-28 shows the locking pliers with the A) jaws closed, B) jaws open using the adjusting screw, and C) jaws open by pushing the movable handle outward.

In its simplest form, a pair of locking pliers employs a one DOF four bar mechanical linkage, similar to that shown in Figure 9-29. As seen in Figure 9-30, when the jaws are closed, the fixed handle and jaw links form a right angle and the movable handle and toggle links are collinear. The jaws can be opened either by moving the movable handle link outward (see Figure 9-31), or by using the adjusting screw, which in effect changes the length of the fixed handle link (see Figure 9-32).

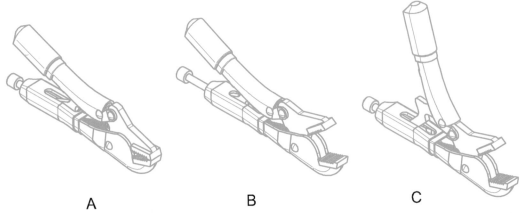

A   B   C

Figure 9-28 Locking pliers in closed and open positions

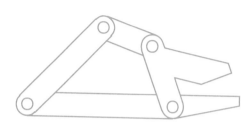

toggle link—

movable handle link

—jaw link

fixed handle link—

A. Jaws Not Shown

B. Jaws Shown

Figure 9-29   Simple four bar linkage model of locking pliers

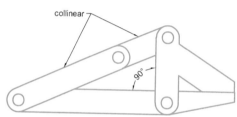

collinear

90°

Figure 9-30   Simple four bar linkage model of locking pliers with jaws closed

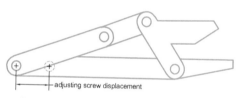

adjusting screw displacement

Figure 9-32   Simple four bar linkage model of locking pliers with jaws open–turn adjusting screw

In fact, the Craftsman locking pliers is more complicated, incorporating as it does a fifth compound link between the movable handle and the toggle link. This compound link serves to make the pliers easier to unlock once clamped, and is at the heart of a patent held for this particular design.

In the event that the CAD software used to create the product's virtual model includes a finite element analysis module, it is possible to estimate the level of stress, strain, or deflection found in a single part, or even an assembly. To do so, it is necessary to model the forces acting on the part, as well as the way in which the part is supported. Figure 9-33A on page 264 shows the loading environment (loads, supports) on a portable water filter handle, assuming that the water filter becomes clogged. Figure 9-33B shows the resulting stress distribution for this loading, based on the von Mises stress criterion.

□ PRODUCT IMPROVEMENT

It can be said that the ultimate goal of reverse engineering is product improvement. In the course of the product dissection process, as the product is used, dissected, documented, and analyzed in various ways in order to understand how it works, ideas for improving upon the existing product often come to light.

Generating improvement ideas is easier for some products than for others. This is largely

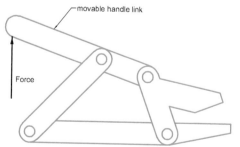

movable handle link

Force

Figure 9-31   Simple four bar linkage model of locking pliers with jaws open–movable handle pushed outward

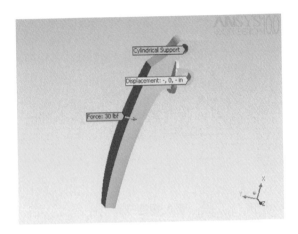

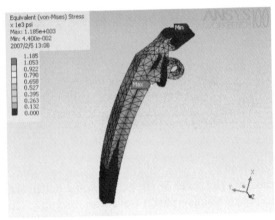

Figure 9-33 Loading and stress pattern on portable water filter handle part

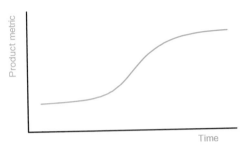

Figure 9-34 S curve showing product innovation over time

influenced by the level of maturity of the product and of the technology on which it is based. Technology and product innovations tend to follow an "S-curve"; in plotting an important measure or metric of the technology versus time, the curve will tend to have an S shape, as seen in Figure 9-34. Early on in the development of a new type of product, innovations are slow to occur, since the market is relatively new. As demand for the product begins to grow, so does the amount of innovation. Finally, as the product reaches maturity, the number of innovations begins to fall off. It is consequently easier to identify product improvement ideas for newer products and technologies.

Locking pliers are an extremely mature product, having been commercially produced since at least 1924. Literally dozens of U.S. patents have been awarded relating to locking pliers inventions, most recently in 2003. The Craftsman design described in this chapter is covered by a U.S. patent that was awarded in 1991.

For this reason, rather than attempting to generate additional product improvement ideas, some innovations in locking pliers design that have taken place over the years will be mentioned. One of the common innovations associated with locking pliers is to vary the jaw configuration in order to create new, more specialized tools. These variations include curved and straight jaw designs, long and bent nose, wrench designs, C-clamp designs, as well as dedicated sheet metal and welding tools. Another popular innovation is the incorporation of a wire cutting tool. By providing cushioned grips, the product's ergonomics is improved, as is its resistance to becoming slippery. Many of the patents associated with locking pliers are related to the development of improved locking and release mechanisms.

Benchmarking is the systematic process of measuring products and services against the toughest competitors. Competitive benchmarking is routinely employed in industry to help identify **_best practices_**. With respect to the locking pliers, the evaluation of competing products can help identify desirable features for adoption.

Finally, the Design for Manufacture and Assembly (DFMA) guidelines, briefly discussed in Chapter 1, offer many potential product improvement ideas. Probably the most important DFMA guideline is to minimize the number of parts. In recent decades the manufacturability and ease of assembly of literally thousands of products have been improved upon through a careful application of these guidelines.

□ REASSEMBLY

Assuming that all components have survived the dissection procedure intact, the product should be reassembled and tested. This next to last step serves to complete the project loop, adding further insight and appreciation for the product's design. The product disassembly plan, exploded view, and assembly animation file can all be used to assist in the product's reassembly.

□ COMMUNICATION

As with any team design project, the product dissection experience offers excellent opportunities for developing both oral and written communication skills. For additional information, see the sections on written reports and oral presentations in Chapter 1.

□ QUESTIONS

PRODUCT DISSECTION PROJECT IDEAS

Bicycle crank set	Drill (electric, manual, power)	Roller blades
Bicycle derailleur	Fabric shaver	Scale (bathroom, food,
Bicycle fork	Fan (oscillating, portable, desk)	postage, triple beam balance)
Bicycle hub	Fire extinguisher	Screwdriver, electric
Bicycle pump (collapsible, telescoping)	Fishing reel	Sprinkler (oscillating, rotary)
Bicycle seat	Flashlight	Stapler
Camera (35mm, disposable)	Gun (paintball, glue, nerf, squirt)	Switch, electrical
Camp stove	Hair dryer	Tape dispenser
Can opener, electric	Hole puncher	Timer
Car (mechanical energy, remote control)	Ice tea brewer	Toothbrush, electric
Car jack	Juicer	Toy (action figure, remote control)
Car visor	Lamp (adjustable, desk, portable)	Transformer toy
Cheese grater, rotary	Mechanical pencil	Whiteout dispenser
Clock (alarm, portable)	Model train	Windshield wiper mechanism
Coffee maker	Outlet, electrical	
Computer mouse	Pencil sharpener	
Corkscrew	Pepper grinder	
Curling iron	Pitching machine, baseball	
Deadbolt lock	Razor scooter	
Door knob	Razor, electric	

WHERE TO LOOK - STORES WHERE APPROPRIATE PRODUCTS, DEVICES, AND MECHANISMS CAN BE FOUND

- Arts and crafts
- Automotive
- Boating
- Cycling
- Department
- Hardware
- Hobby
- Home electronics
- Household, kitchen
- Lawn and garden
- Office supplies
- Outdoor recreation (camping, hiking, climbing, hunting, fishing)
- Sporting goods
- Toy

PERSPECTIVE PROJECTIONS AND SKETCHES

☐ PERSPECTIVE PROJECTION

Historical development

Perhaps the single most important development *in Renaissance art* is the use of perspective. Just prior to this time, paintings like those of Duccio di Buoninsegna (1255–1319) tended to be rather flat and two-dimensional (see Figure A-1). Artists had yet to develop techniques like shading and perspective, as well as an understanding of human anatomy, necessary to create an illusion of depth. Giotto (1267–1337), actually a contemporary of Duccio's, is generally considered to be the first Renaissance painter. In Figure A-2, Giotto employs converging lines to suggest spatial depth, although these lines do not systematically converge to a single vanishing point.

In the work of later Italian Renaissance artists like Leonardo da Vinci (1452–1519) and Raffaello Sanzio (1483–1520) we find paintings that employ one-point perspective to call the attention of the viewer to important details con-

tained in the painting. See, for example, Figures A-3 and A-4. The mathematical rules of perspective were developed and documented by people like the German Dürer (1471–1528) and several Italian artists, including Brunelleschi and Alberti. Filippo Brunelleschi (1377–1446), a Florentine, invented a systematic method for determining perspective projections in the early 1400s. Leon Battista Alberti (1404–1472) wrote the first treatise on perspective, *On Painting*.

Figure A-2   St. Francis Receives Approval of His 'Regula Prima' from Pope Innocent III (1160-1216) in 1210, 1297-99 (fresco), Giotto di Bondone (c.1266-1337)/San Francesco, Upper Church, Assisi, Italy, Giraudon/The Bridgeman Art Library International

Figure A-1   The Maesta, 1308-11 (tempera on panel), Duccio di Buoninsegna, (c.1278-1318)/Museo dell'Opera del Duomo, Siena, Italy/The Bridgeman Art Library International

Figure A-3   The Last Supper, 1495-97 (fresco) (post restoration), Leonardo da Vinci, (1452-1519)/Santa Maria della Grazie, Milan, Italy/The Bridgeman Art Library International

Figure A-4   School of Athens, from the Stanza della Segnatura, 1510-11 (fresco), Raphael (Raffaello Sanzio of Urbino) (1483-1520)/Vatican Museums and Galleries, Vatican City, Italy, Giraudon/The Bridgeman Art Library International

Perspective projection characteristics

As we have already seen in Chapter 3,[1] perspective differs from parallel projection in that in the former the center of projection is a finite distance from the object. The projectors are therefore nonparallel rays that converge to the center of projection. As a consequence, when parallel object edges are not parallel to the projection plane, the edges converge to a *vanishing point* when projected. In addition, objects or features that are further away from the projection plane are more *foreshortened* (i.e., smaller) than closer ones.

The principal advantage of perspective projection is that it produces a more realistic image. It closely approximates the view as seen by the human eye. Conversely, a significant drawback of

perspective projection is that it does a poor job in preserving the shape and scale of the object. Consequently dimensional information often cannot be extracted. In addition, perspective projections are generally more difficult to execute than parallel projections.

Classes of perspective projection

Perspective views are categorized according to the orientation of the object with respect to the projection plane. This orientation determines the number of principal axes (refer again to Figure 3-4 on page 52) that are parallel to the projection plane. If an axis is not parallel to the projection plane, then object edges parallel to this axis will not be parallel when projected. Rather, they will converge to a single point, called a vanishing point. There are three possible cases. These are:

1. One-point perspective (one principal vanishing point)

2. Two-point perspective (two principal vanishing points)

3. Three-point perspective (three principal vanishing points)

In a top view looking down, Figure A-5 illustrates the orientation of three identical cubes (or more generally, three cube-shaped bounding boxes), with respect to a vertical projection plane. Note that, because the scene is viewed from above, the vertical projection plane appears as a line, since it is viewed on edge. Note also that the principal axes of each cube are also represented.

The cube on the left is oriented with one face parallel to the projection plane. If the principal axes of this principal enclosing box (PEB) are extended infinitely in both directions, only one axis intersects the projection plane; the other two axes (one vertical, one horizontal) are parallel to

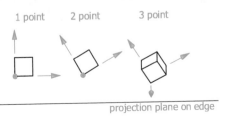

projection plane on edge

Figure A-5   Perspective classes

---

[1] The first part of Chapter 3, up to page 54, serves as a departure point for this discussion of perspective projection.

the projection plane. If the bounding box of an object is oriented in this way, a one-point perspective projection results.

The PEB in the middle has been rotated about a vertical axis so that only the vertical principal axis is parallel to the projection plane; the other two axes are inclined to the projection plane. In this orientation a two-point perspective projection results.

Finally, imagine rotating the middle cube out of the plane of the paper about a horizontal axis. This is the position of the third cube on the right. Notice that, in this case, all three principal axes, when extended, intersect the projection plane. None of the three principal axes are parallel to the projection plane. In this orientation, a three-point perspective projection will result.

Vanishing points

Before discussing these three cases in more detail, it is worth reiterating that, in a perspective projection, if parallel object edges are:

- Parallel to the projection plane, then the projected edges will also be parallel
- Inclined to the projection plane, then the projected edges will not be parallel; they will converge to a *vanishing point*

Referring again to Figure A-5 on page 268, it is apparent that a one-point perspective of a box has one vanishing point, a two-point perspective of a

box has two vanishing points, and a three-point perspective of a box has three vanishing points. These vanishing points are called ***principal vanishing points***, since they are associated with the principal axes of the object. Table A-1 provides a summary of the different perspective classes, along with the number of principal vanishing points of each.

Figure A-6 on the following page illustrates the process of locating principal vanishing points (PVPs) for a two-point perspective projection. Figure A-6B shows the object, projection plane and center of projection (CP) as seen from above. From this we can see the orientation of the object with respect to the projection plane will result in a two-point perspective, since two principal axes are inclined to the projection plane. Two dashed construction lines are drawn through the CP, each parallel to an inclined principal axis, until they intersect the projection plane. Each point of intersection locates a principal vanishing point. Each construction line is parallel to a set of parallel edges on the object, with each edge set parallel to a principal axis. To summarize, taking a line parallel to an inclined object edge and passing it through the center of projection until the line pierces the projection plane locates the vanishing point for that object edge.

Figure A-6C shows the projection plane and the resulting perspective projection. For clarity, the object is not shown. Notice that the principal vanishing points are aligned horizontally, and that the projected edges converge to the PVPs.

Table A-1 Classes of Perspective Projection

Perspective Type	Principal Vanishing Points (PVP)	Principal Axis Orientation
One-point	1	• One principal axis perpendicular to projection plane (PP)
		• Two principal axes parallel to PP
Two-point	2	• Two principal axis inclined to projection plane (PP)
		• One principal axis parallel to PP
Three-point	3	• All three principal axes inclined to PP
		• No principal axes parallel to PP

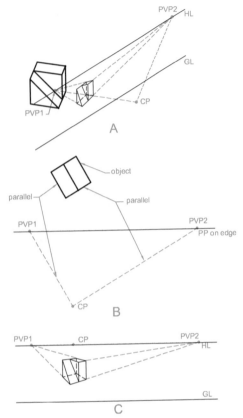

Figure A-6 Locating principal vanishing points

Figures A-6A and A-6C show the ground line (GL) and the horizon line (HL). These lines will be used in later in this Appendix when constructing perspective sketches. The **ground line** is a horizontal line formed by the intersection of the projection plane and the ground plane, i.e. the plane on which the object rests. The **horizon line** represents the eye level of the observer. The horizon line is formed by projecting the CP onto the projection plane, and then passing a horizontal line through it.

Notice that in Figure A-6C both principal vanishing points lie on the horizon line. Also notice that the two sets of actual object edges, parallel to the inclined principal axes, lie in horizontal planes. When projected, these edges converge to the two principal vanishing points. Inclined object edges that lie in parallel planes will converge along the same line when projected.

In the event that these lines lie in horizontal planes, they will converge along the horizon line when projected.

Any parallel group of inclined object edges will converge to a vanishing point in a perspective projection, not just principal axis edges. For example the object depicted in the two-point perspective in Figure A-6C actually has three vanishing points; see Figure A-7 on page 271. The edges of the inclined surface on the actual object are inclined (i.e., not parallel) to the projection plane. Consequently, when projected the inclined surface edges also converge to a vanishing point.

One-point perspective projection

In a one-point perspective projection, one object face is parallel to the projection plane. One principal axis is perpendicular to the projection plane, while the other two principal axes (horizontal, vertical) are parallel to the projection plane.

Figure A-8 on page 271 shows a one-point perspective drawing of a cube. Notice that the vertical edges of the projected cube are parallel to one another, as are the horizontal projected edges. Also notice that the receding[2] edges of the cube are not parallel. Rather, they converge to a principal vanishing point.

Figure A-9 on page 272 shows the perspective arrangement used to obtain Figure A-8. Once again, the top portion of the figure shows a view from above. The object, projection plane, center of projection, projectors, and the construction lines used to locate the PVP are all depicted. Note that the front face of the object is coplanar with the projection plane. The projected image will appear within the encircled area on the projection plane.

The bottom half of Figure A-9 shows the resulting one-point projection (object not shown for clarity). The dotted vertical lines connecting the two portions of the figure are used to locate the projected image on the projection plane.

---

[2] Object edges not parallel to the projection plane will appear to recede back into space when projected.

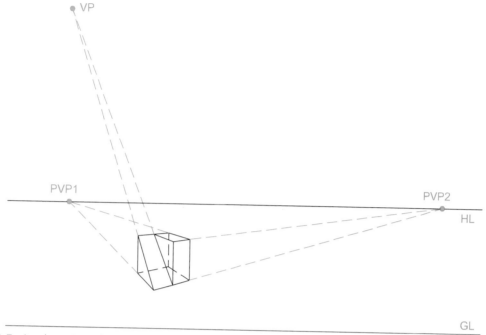

Figure A-7   Another principal vanishing point

Since the front face of the cube lies in the projection plane, it will be projected true size.

Figure A-10 on the following page shows another example of a one-point perspective arrangement, this time with the object entirely behind the projection plane. Notice that, because of this, the front face of the object is not projected true size.

Two-point perspective projection

An example of a two-point perspective arrangement has already been shown in Figure A-6 on page 270. Let us review some of the characteristics:

- One set of principal edges (typically vertical) are parallel to the projection plane, causing the projected edges to also be vertical.

- The other two sets of principal edges, being inclined to the projection plane, will converge to vanishing points when projected. These principal vanishing points will lie on the horizon line.

- If the leading edge of the object lies behind the projection plane (as is the case in Figure A-6), then none of the projected edges will appear true size.

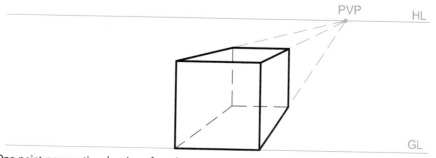

Figure A-8   One-point perspective drawing of a cube

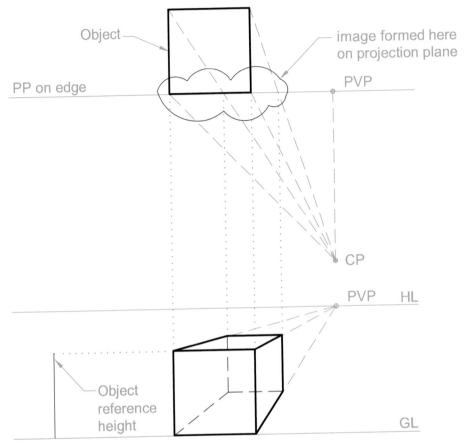

Figure A-9 One-point perspective setup

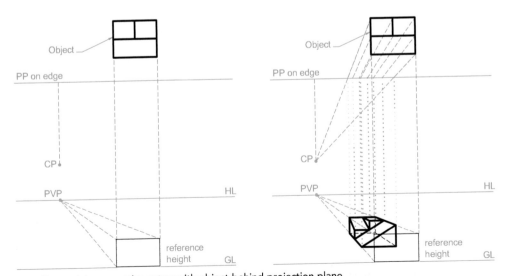

Figure A-10 One-point perspective setup with object behind projection plane

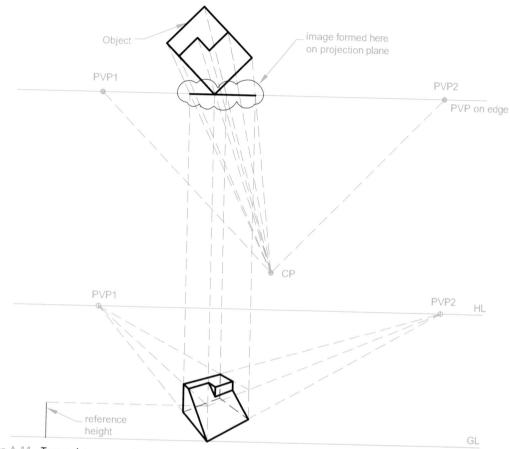

Figure A-11   Two-point perspective projection setup; leading edge lies in projection plane

Figure A-11 provides an example of a two-point perspective projection where the leading object edge lies in the projection plane. In this case, the leading edge is projected true size.

Perspective projection using a 3D CAD system

The procedure[3] described below may be used to create a perspective projection (or, for that matter,

any of the planar projections described in Chapter 3) using a 3D CAD system like AutoCAD®. In this procedure all of the common elements of a projection system (e.g., object, projection plane, center of projection, projectors), as well as the projection itself, are modeled.

1. Start by creating the object, the projection plane, and the center of projection. In Figure A-12 on page 274, the:

   a. Object is modeled as a solid
   b. Projection plane is represented using line segments to draw a vertically oriented rectangle

---

[3] This section is based on the work of Michael H. Pleck, who developed this technique at the University of Illinois at Urbana Champaign.

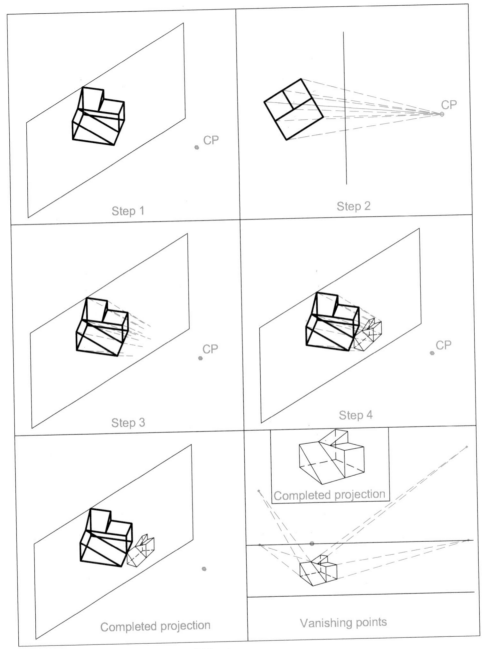

Figure A-12 Perspective projection using a 3D CAD system

c. Center of projection is modeled as a solid sphere or point

2. Use the line command and object snap settings to draw the projectors from the vertices of the object to the CP.

3. Use the trim command to cut the projectors at the projection plane.

4. Use the line command to project each object edge onto the projection plane.

The completed projection is shown in the figure in the lower left corner (projectors not shown). In the lower right corner, projected edges are extended until they meet at vanishing points.

Three-point perspective projection

Because of the difficulty in their construction, three-point perspective drawings are rarely used. Figure A-13 shows an example of a three-point perspective arrangement. Notice that all three principal axes are inclined to the projection plane, and that, when projected, all three sets of edges converge to principal vanishing points. Two of the PVPs lie on the horizon line; one does not.

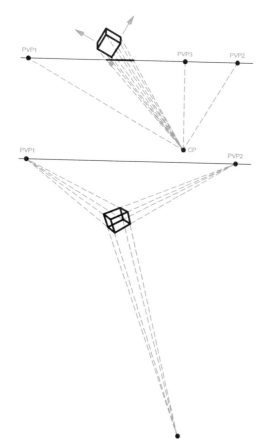

Figure A-13 Three-point perspective setup

Perspective projection variables

There are several variables that influence the appearance of a perspective projection. Some of these variables are discussed below.

PROJECTION PLANE LOCATION

It has already been shown that in a perspective projection, the size of the projected image is dependent upon the location of the projection plane with respect to the object and the center of projection. From Figure A-14 on page 276 (the same as Figure 3-12) it should be clear that the placement of the projection plane affects the size, and even the orientation, of the projected image. The possibilities include that the projection plane:

1. Is behind the object, in which case the projected image is larger than the object
2. Passes through the object, resulting in a projected image that is the same size as the object
3. Is in front of the object, causing the projected image to be smaller than the object
4. Is behind the center of projection. In this case the projected image is inverted.

LATERAL MOVEMENT OF CP

If the center of projection is moved laterally with respect to the projection plane (or equivalently, if the object is moved with respect to the center of projection), different projections will result, as shown in Figure A-15 on page 276. Generally speaking, it is recommended that the center of projection be placed in front of the object, slightly to one side.

VERTICAL MOVEMENT OF CP

Figure A-16 on page 276 shows how the projection of an object can change, depending upon the vertical placement of the center of projection with respect to the ground plane. In Figure A-16A on the left, the center of projection is above

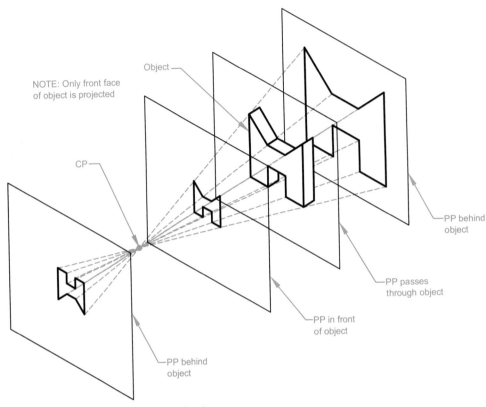

Figure A-14 Perspective projection plane projection

the object. Figure A-16B in the middle shows the same object, but with the center of projection at

the same level as the object. Finally, in Figure A-16C on the right, the center of projection is below the object.

VARYING DISTANCE FROM CP

One of the strengths of perspective projection is that it results in a more realistic image than parallel projection. This is due to the fact that,

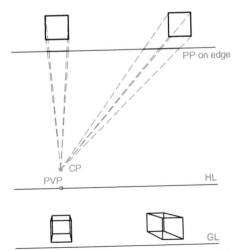

Figure A-15 Lateral movement of an object with the same center of projection

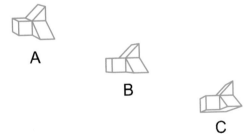

Figure A-16 Vertical movement of the center of projection

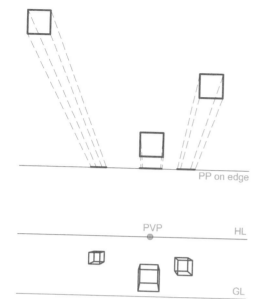

Figure A-17 Varying object distance from projection plane

much like our own vision, the size of an image projected using perspective depends upon the distance of the object to the projection plane. In Figure A-17 the same size cube is projected using one-point perspective. Notice that the further the object is from the projection plane, the smaller the projected image.

□ PERSPECTIVE SKETCHES

Introduction

Perspective sketches provide a more realistic representation of an object than parallel projection techniques, while sacrificing much of the latter's ability to preserve dimensional information. Perspective sketches are also more difficult to construct than either oblique or isometric sketches.

A perspective sketch represents an object as an observer would see it from a certain vantage point. Receding parallel object edges converge in a perspective pictorial, causing distant objects to appear smaller. In contrast, in a parallel projection, parallel edges remain parallel in the projected image. Using parallel projection,

objects are projected as the same size, regardless of their distance from the projection plane.

Terminology

Key elements of a perspective sketch are shown in Figure A-18 on the following page. As we have already seen, these elements include the ground line, the horizon line, and vanishing points. The ground line represents the plane on which the object rests, and is formed by the intersection of the ground plane with the projection plane. The horizon line represents the eye level of the observer.[4] A vanishing point is a position on the horizon to which depth projectors converge.[5]

One-point perspective sketches

Recall that in a one-point perspective projection, one object face is parallel to the projection plane. This explains the similarity between a one-point perspective and an oblique sketch, which is also oriented with two principal axes parallel to the projection plane. See Figure A-19 on the following page for a comparison between one-point perspective and oblique pictorials.

The main difference between the two pictorials is that the receding edges are parallel in oblique projection, whereas in the one-point perspective the receding edges converge to a vanishing point.

In a one-point perspective projection, if the object's front face coincides with the projection plane, then the front face of the object is projected full scale (see Figure A-14). Otherwise, if it lies behind (or in front of) the projection plane, the projected front face will be smaller (or larger) than the actual. In practical terms, though, when constructing a perspective sketch, a vertical edge

---

[4] Recall from the discussion of perspective projection earlier in the chapter that the horizon line is at the same height as the center of projection.

[5] In a perspective projection, parallel object edges inclined to the projection plane converge when projected. If these edges are also parallel to the horizontal ground plane, they will be projected along the horizon line.

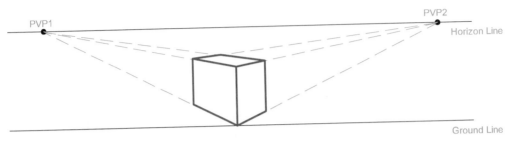

Figure A-18 Perspective sketch terminology: two-point perspective

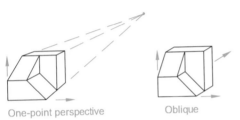

One-point perspective   Oblique

Figure A-19 Comparison of one-point perspective and oblique sketches

length representing the height of the front face is first chosen. The horizontal width dimension is then scaled proportional to this vertical dimension.

Scaling along the receding depth axis involves some visual approximation. Figure A-20 shows two one-point perspective sketches of the same cube. On the left, the cube has been laid out using the same distance L along the horizontal, vertical and converging axes. Clearly, the resulting pictorial appears to be too long in the receding axis direction. This distortion is because foreshortening along the depth axis has not been accounted for. On the right the depth dimension has been reduced, resulting in an improved representation

of a cube. Although the amount of foreshortening depends on several variables, a rule of thumb is to foreshorten the converging axis dimension on a one-point perspective sketch to approximately two-thirds of the actual.

Two-point perspective sketches

In a two-point perspective sketch, the object is oriented so that only one principal axis, typically vertical, is parallel to the projection plane. The other two principal axes are inclined to the projection plane. As a consequence, vertical object edges remain parallel when projected, while the two other sets of principal edges converge to different vanishing points. Both of these principal vanishing points lie on the horizon line (see Figure A-21).

If the projection plane passes through the leading vertical edge of the object, this edge will be projected full size (refer to Figure A-14 on page 276). Otherwise, if the edge is behind (or in front of) the projection plane, the projected vertical will be smaller (or larger) than the true length. When constructing a two-point perspective sketch, though, a vertical edge length representing the height of the leading edge of the bounding box is simply chosen, without regard for the location of this edge with respect to the projection plane.

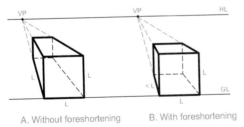

A. Without foreshortening   B. With foreshortening

Figure A-20 One-point perspective of a cube, without and with foreshortening

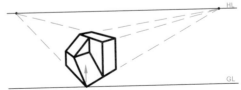

Figure A-21 Two-point perspective sketch

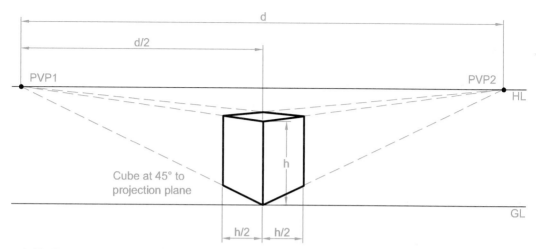

Figure A-22   Two-point perspective pictorial of a cube at 45-degree angle to projection plane

Convergence lines are then drawn from the leading edge endpoints to both principal vanishing points.

As was the case with one-point perspective, the amount of foreshortening along the receding axis must be estimated in order to create a well-proportioned sketch. In the case of a two-point perspective, however, there are two receding axes.

Figure A-22[6] shows a two-point perspective sketch of a cube, where the cube is placed at a 45-degree angle to the projection plane. If the horizontal distance from the leading edge to each principal vanishing point is equal, as is the case in Figure A-22, then the amount of foreshortening along each receding axis will be equal. A good estimate of this foreshortening amount is provided in the figure.

Another scenario is provided in Figure A-23. In this case the cube is positioned at a 30-degree angle with respect to the projection plane. The cube is laterally positioned so that its leading edge is one-fourth the distance between the two

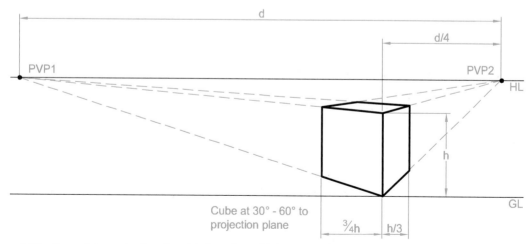

Figure A-23   Two-point perspective pictorial of a cube at 30-degree angle to projection plane

[6] Figures A-22 and A-23 are taken from *Graphics for Engineers,* 2nd Edition, by Jerry Dobrovolny and David O'Bryant, John Wiley & Sons, 1984.

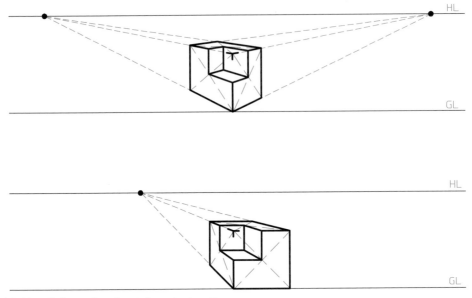

Figure A-24 Use of diagonals to locate important vertices

principal vanishing points. Given this scenario, Figure A-23 provides reasonable foreshortening estimates along both receding axes. Note that the closer the PVP to leading edge distance, the greater the amount of foreshortening.

Proportioning Techniques

A useful proportioning technique when constructing perspective sketches is to sketch the diagonals of a receding face in order to locate the midpoint of that face. This point can then be projected to an adjacent edge in order to locate other key vertices. Figure A-24 illustrates this technique for both one and two-point perspective sketches.

In Figure A-25 this technique is extended to allow partitioning of a trapezoidal area into thirds and quarters. See the section on partitioning lines in Chapter 2 for additional information.

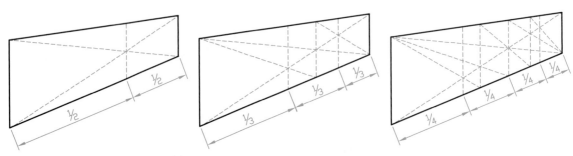

Figure A-25 Partitioning trapezoidal areas

Step-by-step one-point perspective sketch example

Given a cavalier oblique pictorial (PEB is a cube), a reference edge height and location, and a principal vanishing point, construct a one-point perspective sketch of the object. See Figure A-26.

1. Use construction lines to complete front face of bounding box.

2. Use construction lines to sketch convergence lines.

3. Estimate the foreshortened depth, and use construction lines to complete the bounding box.

4. Using construction lines, sketch diagonals on receding (top, right) faces.

5. Using construction lines, locate key features. Midpoints of horizontal and vertical edges are used to locate 2, 5, and 7. Intersecting diagonals used to locate mid-face vertices (1, 3). Remaining vertices (4, 6) located by passing horizontal and/or vertical lines through existing vertices to find intersections.

6. Go bold.

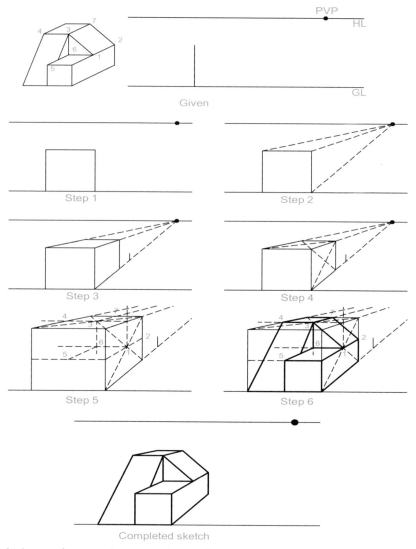

Figure A-26   Multiple steps for one-point perspective sketch

Given a cavalier oblique, the reference edge height and location, and the location of the principal vanishing points:

1. Use construction lines to sketch convergence lines to PVP1 and PVP2. Also lay out the unforeshortened dimensions of the object's PEB.

2. After estimating the foreshortened depths, complete the bounding box (use construction lines).

3. On the front face sketch the diagonals of the face. Also pass a vertical line through the point of intersection of the diagonals.

4. In order to partition the front face into three segments, sketch the diagonal lines shown in step 4.

5. Sketch two more vertical lines on the front face, each one passing through the intersection formed by the diagonal lines created in steps 3 and 4.

6. Sketch the diagonals on the left and top faces.

7. Sketch a vertical line from the intersection of the left face diagonal to the upper left edge of the bounding box, and then sketch a line from this intersection point to the intersection of the top face diagonals. Finally, extend this line until it intersects the right edge of the top face.

8. Go bold.

The completed sketch is shown in Figure A-27.

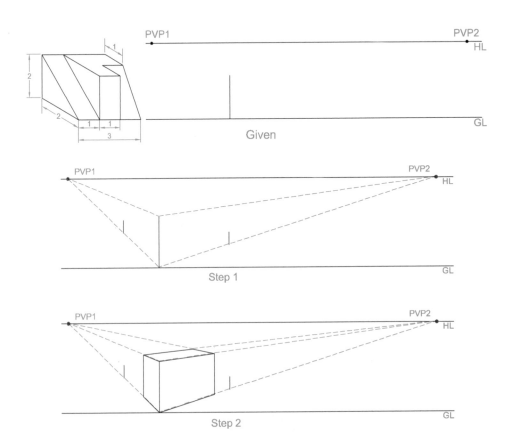

Figure A-27 Multiple steps for two-point perspective sketch

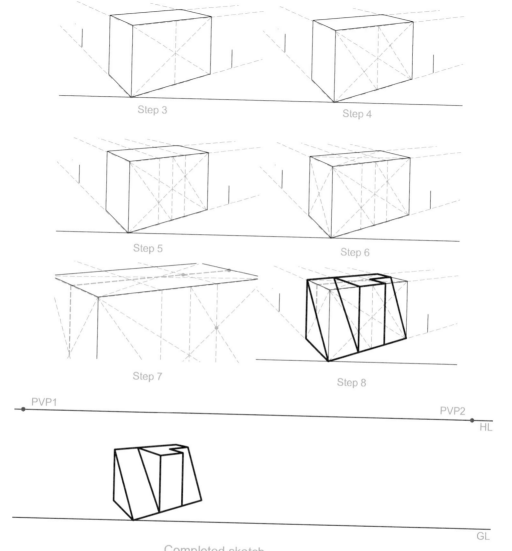

Step 3

Step 4

Step 5

Step 6

Step 7

Step 8

PVP1

PVP2
HL

Completed sketch

Figure A-27 (Continued)

GL

Summary: Orientation of pictorial sketching axes

A. Oblique
- All 3 sets of PEB edges (horizontal, vertical, receding) remain parallel

B. Isometric
- All 3 sets of PEB edges (vertical, 30° to right, 30° to left) remain parallel

C. One-point perspective
- 2 sets of PEB edges (horizontal, vertical) remain parallel
- 1 set converges to PVP

D. Two-point perspective
- 1 sets of PEB edges (vertical) remain parallel
- 2 sets converge to PVP's

See Figure A-28.

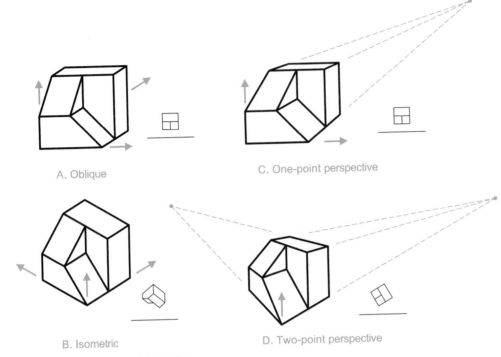

A. Oblique

C. One-point perspective

B. Isometric

D. Two-point perspective

Figure A-28 Orientation of pictorial sketching axes

☐ QUESTIONS

TRUE AND FALSE

1. A two-point perspective has two principal axes parallel to the projection plane.

2. Two-point perspective and oblique projections have the same number of principal axes inclined to the projection plane.

3. In a perspective projection, if the projection plane is located in front of the object, the projected image will be smaller than the object.

MULTIPLE CHOICE

4. Figure PA-1 shows a perspective view of a vertical pole projected onto a projection plane. If the length of the pole is 30 feet, what is the approximate height of the observer? (i.e., the distance from the ground to the observer's eye level) :

a. 0 feet

b. 3 feet

c. 5 feet

d. 10 feet

e. 15 feet

f. 30 feet

g. Not Determinable

SKETCHING

5. Given the isometric view of the cut block objects appearing in P3-4 through P3-65, use DwgA-1 (or download worksheet from the book website) to sketch a one-point perspective view of the object.

6. Given the isometric view of the cut block objects appearing in P3-4 through P3-65, use DwgA-2 (or download worksheet from the book website) to sketch a two-point perspective view of the object.

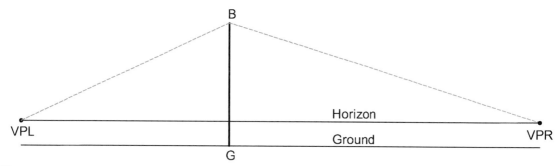

Figure PA-1 (Figure adapted from the work of Michael H. Pleck)

7. Given the cavalier oblique view of the cut block objects appearing in P3-65 through P3-95, use DwgA-1 (or download worksheet from the book website) to sketch a one-point perspective view of the object.

8. Given the cavalier oblique view of the cut block objects appearing in P3-65 through P3-95, use DwgA-2 (or download worksheet from the book website) to sketch a two-point perspective view of the object.

B GEOMETRIC DIMENSIONING AND TOLERANCING

☐ INTRODUCTION

Recall from Chapter 6, Dimensioning and Tolerancing, that variation from perfect geometry is inevitable in all manufactured parts. The tolerancing techniques presented in Chapter 6 serve to limit the extent to which this variation is permitted. In using these conventional tolerancing methods on high precision parts, however, the only alternative left to the designer to ensure a proper fit is to tighten these tolerances. This approach leads to parts that are both expensive and difficult to produce.

Geometric Dimensioning and Tolerancing (GD&T) is an internationally accepted, symbol-based system of tolerance controls. GD&T consists of a well-defined set of symbols, rules, definitions, and conventions that can be used to control not only the size, but also the form, location, orientation, profile, and runout of part features. This system is fully documented in ASME Y14.5M-1994, Dimensioning and Tolerancing. ASME Y14.5M is more than 95% compatible with International Organization for Standardization (ISO) standards.

Many GD&T techniques were developed and used in World War II to help mass produce ships, aircraft, and ground vehicles. In recent decades, geometric tolerancing has experienced resurgence in interest, owing in part to the widespread acceptance of the ISO 9000 global quality standards. In addition to being supported by both national and international standards, GD&T offers a number of other advantages to companies engaged in cutting-edge manufacturing. These advantages include increased tolerance zone areas, fewer assembly problems, a datum system that ensures repeatable measurements and improved communications, a tolerance system that takes part function into account, and a stable platform for the successful use of statistical quality control (SPC).

☐ JUSTIFICATION

Conventional tolerances (also called coordinate or traditional tolerances) control the variability of linear dimensions that describe the size and location of part features. They do not, however, address the variability of a part feature's shape. Conventional tolerancing techniques are unable to control all aspects of part shape, aspects such as straightness, flatness, circularity, cylindricity, parallelism, and perpendicularity. Further, conventional tolerances do not explicitly use the concept of datums. Ordered datums are important to both manufacturing and inspection, providing a consistent frame of reference for the measurement of part features. Another weakness of traditional tolerance methods is that location control is cumbersome. Using conventional tolerances to locate the center of a hole, for example, results in an inefficient rectangular tolerance zone, as we will soon see. Geometric tolerancing was developed in large part to address these shortcomings. GD&T provides part shape controls, explicit datum definition and precedence, and improved location control.

Conventional tolerances, while easy to understand and use, are often ambiguous, and do not allow for the control of form. Geometric tolerances, on the other hand, are harder to understand and use, but are more specific, and allow for the control of both location and form. As a consequence, conventional tolerances are still used in many manufacturing facilities where the tolerance requirements are less demanding.

While universal acceptance of GD&T is still a long way off, as CAD modeling of parts continues to promote geometric complexity, and as part tolerances require more precision, geometric tolerancing will continue to gain acceptance. Currently, geometric dimensioning and tolerancing is the only method to define mechanical parts so that they fit and function with the widest, cheapest tolerances possible.

□ SIZING AND POSITIONING OF HOLES: PLATE WITH HOLE EXAMPLE

One of the most powerful uses of GD&T is to accurately locate holes. In this section a practical example of how GD&T handles this task is discussed. Shown in Figure B-1 is a part, an irregularly shaped plate with a hole. Prior to defining the position of the hole, we must first establish three ordered mutually perpendicular datum planes, called a ***datum reference frame***.

Datum reference frame

Many different groups (e.g., design, fabrication, inspection) need to measure a part in the course of its manufacture. It is consequently important to establish a standard method for fixturing parts for measurement, so that these measurements are repeatable. The part is first placed on a smooth horizontal surface, as shown in Figure B-2. This surface is labeled datum A. However flat and smooth the bottom surface of the plate may be, it will still not be perfectly flat. The part will consequently be in contact with datum A at a minimum of three points (three points determine a plane) along its bottom surface. Recall from the degrees of freedom (DOF) discussion in Chapter 7 that a rigid body has six degrees of freedom. Placing (mating) the part on (with) the datum surface constrains three DOF's, with three DOF's still remaining (one in rotation, two in translation).

Figure B-3 shows an additional datum surface labeled B. Datum B is perpendicular to datum A. By sliding the part on datum A until it comes firmly into contact with datum B at a minimum of two points, the part is further constrained (see Figure B-4). The part now comes into contact with datum B at two high points along its back surface, while also contacting datum A at three high points along its bottom surface. The part's movement is now constrained so that it can only translate along the length of datum B. Five DOF's have been constrained, with one DOF remaining.

Figure B-1   Plate with a hole

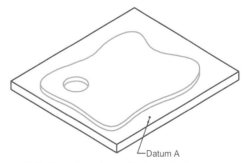

Figure B-2   Part placed on a horizontal (datum A) surface

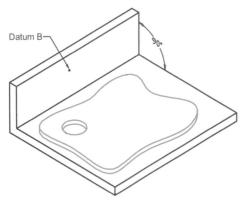

Figure B-3   Datum B surface added, perpendicular to datum A

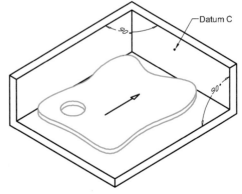

Figure B-5   Datum C surface added, mutually perpendicular to A and B

Figure B-5 adds a third datum (C), mutually perpendicular to A and B. As the part slides along A and B, it will come into contact with C at a single high point, and is therefore fully constrained. The part is ready for measurement.

The order in which the part is brought into contact with the datums is important. If, for example, the part is first positioned with respect to datum C, and then datum B, the part would be positioned differently, as shown in Figure B-6B on page 289.

Although we have been discussing an irregularly shaped plate, the same logic applies to a plate with a rectangular shape. No matter how precise the plate's fabrication, variability in the form of surface roughness, out-of-perpendicular

corners, etc., requires a consistent fixturing approach that ordered datums can provide.

Hole positioning

Now that the part has been fixtured so that its position is uniquely defined and can easily be replicated by others, we can move to the problem of locating the hole. Figure B-7 on page 289 shows two dimensions indicating the position of the hole with respect to datums B and C. Conventional plus and minus tolerances have been applied to these dimensions. As a consequence of these tolerances, the resulting tolerance zone within which the center of the hole must lie (also shown in Figure B-7) is a 10 × 10 square.

Whenever conventional coordinate tolerances are used to locate the center of a hole, the result is a rectangular tolerance zone. Figure B-8 on page 289 shows that by replacing this area with a circle whose diameter is equivalent to the diagonal of the rectangle, a 57% increase in usable tolerance (shaded areas) results. The small crosses represent hole centers that would be acceptable assuming a circular tolerance zone, but that are rejected using coordinate location tolerances.

Now take a look at Figure B-9 on page 290, which shows the equivalent situation using GD&T. Notice that the linear dimensions are untoleranced, and that they are box framed. These are called *basic dimensions*, and represent

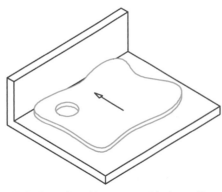

Figure B-4   Part placed in contact with datum B

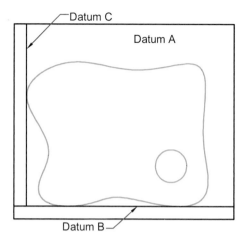

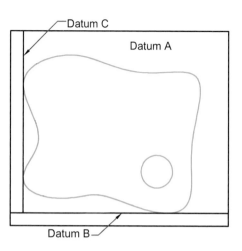

A. Orginal - Datum B first, then C

B. Datum C first, then B

Figure B-6 Alternative placement; first datum C, then B

the exact or true position of the circular tolerance zone (also shown in Figure B-9) within which the hole center must be located. The diameter of the circular tolerance zone is indicated in the *feature control frame*, shown below the ∅50 ±5 hole diameter dimension. The feature control frame is divided into compartments. The first compartment contains the geometric characteristic symbol for position, ⊕. The second compartment contains the *positional tolerance* for the hole, ∅10. Succeeding compartments show the datums, A, B, and C. The ±5 tolerance on the ∅50 hole diameter indicates that the hole diameter may vary between

45 and 55, but the center of the hole must still lie within the ∅10 diameter circular tolerance zone.

In fact this tolerance zone is actually a cylinder with a diameter of 10. The basic dimensions locate the central axis of this tolerance zone, with the axis parallel to both B and C, and perpendicular to A. The central axis of the actual hole must lie entirely within the cylindrical tolerance zone.

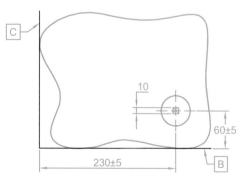

Figure B-7 Conventional plus and minus tolerances used to locate hole

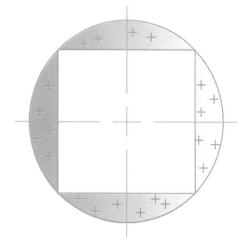

Figure B-8 Hole location tolerance zone; circular versus rectangular

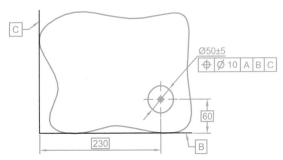

Figure B-9 Geometric dimensioning of hole location

Material condition

Maximum material condition (MMC) refers to the condition in which an external feature (e.g., a shaft) is at its largest allowable size, or that an internal feature (e.g., a hole) is at its smallest possible size. For the hole feature of the part in question, the MMC is ∅45.

The opposite of MMC is *least material condition*, or LMC. For an external feature like a shaft, LMC is the smallest possible size. For an internal hole feature, LMC is the largest allowable size.

Virtual condition

The worst possible condition for a mating part in an assembly occurs when the feature of size (i.e., the hole) is at MMC and is positioned at the extreme limits of the location tolerance zone. This is referred to as the hole feature's *virtual condition*. Figure B-10A shows the hole feature in virtual condition.

For a hole, the virtual condition can be calculated as follows:

Hole virtual condition = MMC diameter − location tolerance zone diameter = ∅45 − ∅10 = ∅35

A feature's virtual condition can be graphically represented as a constant theoretical boundary located at the feature's true position; see Figure B-10B. For an internal feature like a hole to be acceptable, no hole surface elements can lie within this boundary. Notice that in Figure B-10B the worst case hole (MMC, positioned at extreme limits of location tolerance zone) does not lie inside the virtual condition boundary. For an external feature like a shaft, no shaft surface elements can lie outside the boundary.

A *functional gage* is a fixed gage (i.e., one with no moving parts) that simulates the worst possible condition of a part to be mated with a part under inspection. Functional gages are created with dimensions equal to the virtual condition of a part in order to simulate the fit of the mating parts.

Bonus tolerance

Figure B-11 on page 291 is the same as Figure B-9, but with an additional symbol in the feature control frame. The M with a circle around it stands for maximum material condition (MMC). The MMC modifier means that if the hole departs from MMC (∅45), then the difference between the actual hole

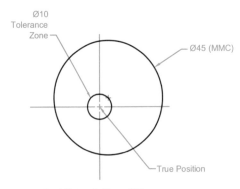

A. Virtual Condition   B. Virtual Condition Boundary

Figure B-10 Hole virtual condition

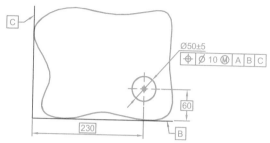

Figure B-11   Geometric dimensioning of hole, MMC modifier

size and MMC can be applied to the hole location tolerance, as shown in Table B-1 below.

This increase in position tolerance, due to the feature's departure from MMC, is called a **bonus tolerance**. Bonus tolerance comes from the feature of size tolerance (as the feature departs from MMC), and is applied to the feature location tolerance. The bonus tolerance can be graphically represented as an area between two circles, as shown in Figure B-12. The fixed inner diameter is the positional tolerance at MMC ($\varnothing$10), while a variable outer diameter includes this positional tolerance, plus any added tolerance as the hole departs from MMC. Figure B-12 shows the outer extent of the bonus tolerance when the hole size is at LMC, ($\varnothing$20). Notice that both location tolerance zone circles are located at the true position of the hole center. Also shown is an LMC ($\varnothing$55) hole, positioned at an extreme limit of its locational tolerance zone.

Finally, Figure B-13 shows both the bonus tolerance and the virtual condition for the hole feature. Notice that the worst case LMC ($\varnothing$55) hole is entirely outside the virtual condition boundary, indicating that the hole is within tolerance.

Table B-1

Actual hole size	Added tolerance	Positional tolerance	Combined tolerance
45 MMC	0	10	**10**
48	3	10	13
50	5	10	15
52	7	10	17
55 LMC	10	10	**20**

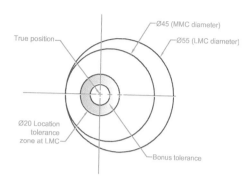

Figure B-12   Bonus tolerance representation

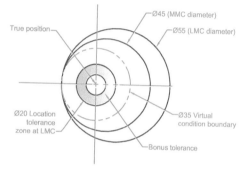

Figure B-13   Bonus tolerance and virtual condition

☐ DATUMS

A **datum** is a theoretically exact point, axis, or plane. Datum elements exist within the framework of the datum reference frame, three mutually perpendicular intersecting planes established by physical contact with datum features on the part. See Figure B-14. A **datum feature**, in turn, is an actual part feature used to establish a datum. A **datum simulator** is a datum found on inspection or processing equipment that is used to simulate a theoretical datum. Since the datum reference frame only exists in theory, it must be represented by datum simulators.

As we have already described in the previous section, datum features must be designated in an order of precedence in order to properly position the part with respect to a datum reference frame. Assuming that the datum features are all plane surfaces, the primary datum feature is brought into contact with the first datum plane, where

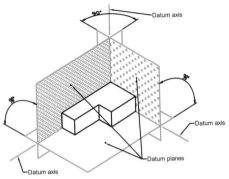

Figure B-14   Datum reference frame

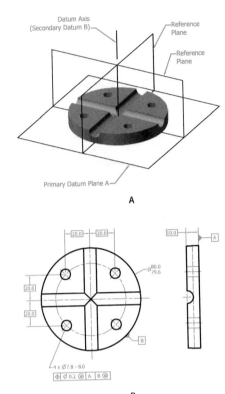

Figure B-16   Part with cylindrical datum feature

contact is made at a minimum of three points. At least two points on the secondary datum feature are then placed in contact with the second datum plane. Finally, at least one point on the tertiary datum feature is brought into contact with the third datum plane, thus completing the process. Using this procedure for positioning the part on the datum reference plane ensures a common basis for measurements.

As seen in Figure B-15, datum feature reference letters are used to identify the datum features. The order of precedence of the datum features is established, from left to right, in the feature control frame.

Cylindrical features can also be used as datums. As shown in Figure B-16A, the cylindrical feature datum B is associated with two orthogonal datum reference planes. In a circular view (see Figure B-16B, front view), these reference planes are represented by the perpendicular centerlines of the circular feature. Typically there is another datum plane, orthogonal to the cylindrical datum (i.e., datum plane A in Figure B-16).

In the event that a part's surface is not appropriate for use as a datum feature, either because it is non-planar or irregular (e.g., castings, forgings), then it becomes necessary to use other geometry to precisely position the part. *Datum targets* are specified points, lines, or areas on a part that are identified with basic dimensions in order to establish datums. Figure B-17 on page 293 shows a part drawing in which the datum D is defined by the datum target point D1 and the datum target line D2.

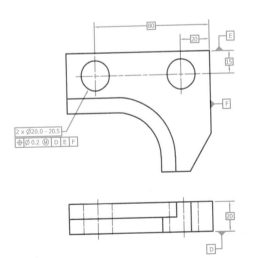

Figure B-15   Datum feature reference letters

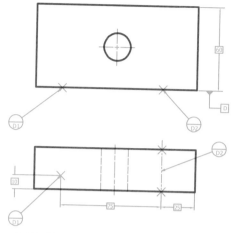

Figure B-17 Datum targets

□ MATERIAL CONDITION

We have already seen that maximum material condition (MMC) indicates that the feature contains the maximum amount of material permitted for that feature. External features (e.g., shafts, pads, and bosses) are at MMC when at maximum size, while internal features (e.g., holes, slots, pockets) are at MMC when at minimum size.

A less common term related to feature size is least material condition (LMC). A feature at LMC contains the least amount of material permitted by the toleranced dimensions; LMC for a shaft is the minimum size; for a hole LMC indicates the largest size that is still within tolerance.

Additionally, the term regardless of feature size (RFS), indicates that the tolerances apply to a geometric feature regardless of its size, ranging between MMC and LMC.

□ GD&T RULE 1

Rule 1, Individual Features of Size, from ASME Y14.5M-1994 states that when only a tolerance of size is specified, the limits of size for this feature shall be used to determine the extent to which variations in the feature's geometric form, as well as its size, are allowed. In effect, Rule 1 points out that all dimensions have built-in form (e.g.,

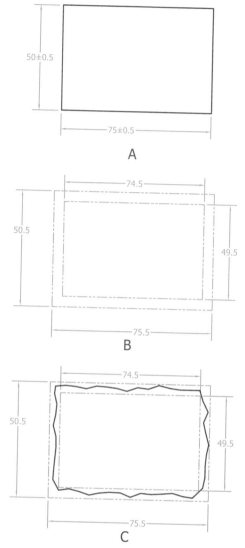

Figure B-18 Envelope rule

straightness, flatness, roundness) controls, and that additional geometrical controls are only required when size and position controls need refinement. An important consequence of Rule 1 is that GD&T controls should only be used when conventional tolerances do not provide the desired level of control over a feature's form.

In way of illustration, Figure B-18A shows a rectangular feature whose size has been defined using conventional plus and minus tolerances. Figure B-18B displays the upper and lower

Type of Feature	Type of Tolerance	Characteristic	Symbol	Comments
Individual	FORM	Straightness	—	
		Flatness	⬦	
		Circularity (Roundness)	O	
		Cylindricity	⌀	
Individual or Related	PROFILE	Profile of a line	⌒	
		Profile of a surface	⌓	
Related	ORIENTATION	Angularity	∠	
		Perpendicularity	⊥	
		Parallelism	//	
	LOCATION	Position	⊕	commonly used
		Concentricity	◎	hard to inspect
		Symmetry	⩶	hard to inspect
	RUNOUT	Circular runout	↗	
		Total runout	↗↗	

Figure B-19   Geometric characteristic symbols

boundaries, or tolerance zone, within which all elements of the rectangular feature must lie, based upon these bilateral tolerances. Finally, Figure B-18C shows one possible shape that conforms to the specified size tolerance limits. From this it can be seen that size limits control the extent to which geometric form may vary. A further point is that, should a form tolerance also be applied to the feature, the form tolerance zone would then need to be smaller than, and fall within the size limits. In this way the form tolerance serves as a refinement of the size tolerance.

☐ GEOMETRIC TOLERANCE SYMBOLS

The conventional (or coordinate) tolerances discussed in Chapter 6 are used to control feature size (i.e., length, diameter), and to a large degree, feature location. Earlier in this chapter an example using geometric tolerancing methods to control the position of a hole center was discussed. Figure B-19, adapted from ASME Y14.5M-1994, shows the geometric characteristic tolerance symbols, organized in the following categories: form, profile, orientation, location, and runout.

Figure B-20 shows datum feature symbols applied to a feature surface (A, B) or to a feature of size (axis (C, E), center plane (D)). Basic dimensions, used to describe the theoretically exact size or location of a feature, are represented as shown in Figure B-21 on page 295. Maximum and least material condition modifying symbols are shown in Figure B-22 on page 295.

Geometric tolerances are expressed in a compartmentalized feature control frame, which

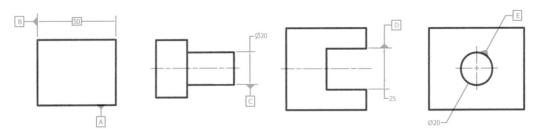

Figure B-20   Datum feature symbols

Figure B-21   Basic dimensions

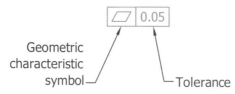

Figure B-22   MMC and LMC modifying symbols

Figure B-23   Feature control frame

includes at minimum the geometric characteristic to be controlled and a tolerance value (see Figure B-23). As shown in Figure B-24, modifying symbols may also be represented in the feature control frame, as well as datum reference letters (see Figure B-25).

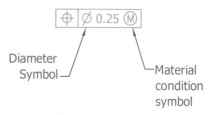

Figure B-24   Modifying symbol in feature control frame

Figure B-25   Datum reference letters in feature control frame

Figure B-26 shows a detail drawing of a cylindrical part using geometric tolerancing techniques.

□ FORM TOLERANCES

Geometric form tolerances are used to control variability in the shape of individual features such as surfaces. Form tolerances include straightness,

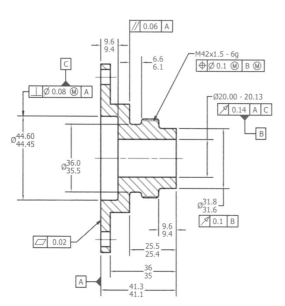

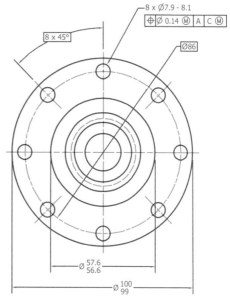

Figure B-26   Geometrically dimensioned cylindrical part (taken from ASME Y14.5M-1994)

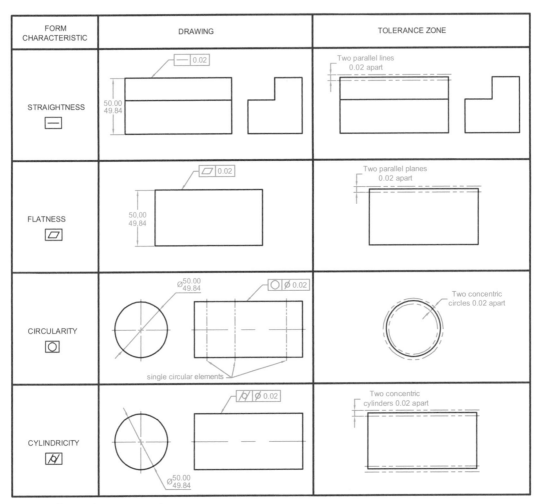

Figure B-27   Form tolerance examples

flatness, circularity (or roundness), and cylindricity. Since form tolerances only apply to individual features (not being related to other features), they are not associated with datums. As shown in Figure B-27, form tolerances establish a tolerance zone which is defined by either a linear or radial distance between two parallel or concentric boundaries. A form tolerance applied to an individual surface feature must be smaller than, or a refinement of, the size tolerance.

Straightness can also be applied to a *feature of size* by associating the form control with a size dimension. Feature of size examples include the diameter of a cylindrical pin, or the distance between two parallel surfaces. Figure B-28 on the following page provides examples of the

straightness control applied to both an individual feature (i.e., a cylindrical surface) and to a feature of size (i.e., diameter of a cylinder). Notice that when applied to a size dimension, the straightness tolerance controls the axis (or median plane) of the feature, and not the external surface. In this case the straightness control may be modified to apply at MMC, thus providing a bonus tolerance as departure from MMC occurs.

☐ PROFILE TOLERANCES

Profile tolerances are used to control the shape of irregular features. Prior to using profile tolerancing, a true profile must be established

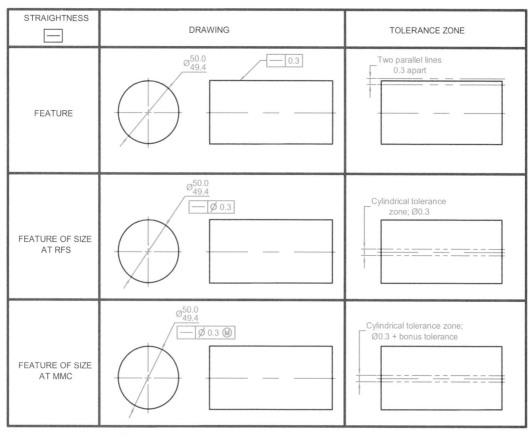

Figure B-28 Straightness tolerance examples

using basic dimensions, and may consist of line segments, arcs, and other curves. The profile tolerance zone consists of a uniform boundary along the true profile within which the elements of the surface must lie. As shown in Figure B-29, a profile tolerance can either apply to cross sections through the part (profile of a line), or to an entire surface (profile of a surface),

□ ORIENTATION TOLERANCES

Of the geometric characteristic tolerance controls, three are used to establish specific orientation relationships between related features. They are angularity, parallelism, and perpendicularity. Since they involve feature relationships, orientation tolerances always use a datum reference. In addition to providing specific controls

between related features, orientation tolerances also provide an indirect control on the form of an individual feature. For example, parallelism can be used to control axis straightness at some distance from a datum feature. Since parallelism is referenced to a datum, it is more restrictive than either straightness or flatness.

Although the orientation control examples shown in Figure B-30 are applied to individual (surface) features, orientation controls can also be applied to features of size. When used in this way angularity, perpendicularity, and parallelism can all be used to establish a cylindrical tolerance zone within which the axis of a cylindrical feature of size must lie. As with the straightness control (see the last example in Figure B-28), the orientation controls may then be modified to apply at MMC, thus providing a bonus tolerance as departure from MMC occurs.

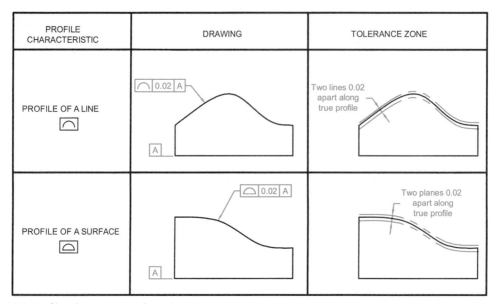

Figure B-29   Profile tolerance examples

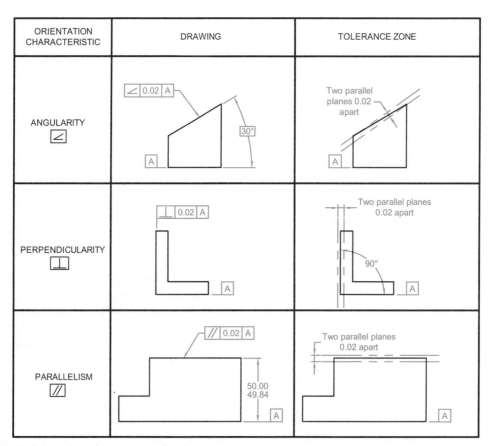

Figure B-30   Orientation tolerance examples

☐ LOCATION TOLERANCES

The geometric tolerances position, concentricity, and symmetry are all used to locate geometric features with respect to a datum. In most situations, though, position tolerances can be used to control feature symmetry and concentricity, at significantly reduced cost. Owing to its flexibility and major cost advantages, position tolerancing is used more than any other geometric tolerance control.

Position tolerancing is so effective because it eliminates the inefficient rectangular tolerance zone problems inherent to conventional tolerancing techniques, while taking advantage of MMC bonus tolerance opportunities. An important outcome of positional tolerancing is that functional (fixed) gages can be used to verify the conformance of hole features. Finally, position tolerancing and the accompanying basic dimensioning techniques that it is based upon can be used to avoid tolerance accumulation.

As seen in the earlier section covering the positioning of a hole in a plate, conventionally toleranced location dimensions result in a rectangular tolerance zone for locating the center of a hole. Positional tolerances, on the other hand, result in a more efficient circular tolerance zone centered at the true position of the hole. To accomplish this, untoleranced basic dimensions are used, referenced with respect to datums, to represent the theoretically exact position of the hole. This circular tolerance zone is specified in a feature control frame.

It is common practice to assume that a positional tolerance applies at MMC. In so doing, an increased positional tolerance becomes available as the feature departs from MMC, towards LMC. This increase in the positional tolerance zone is the previously discussed bonus tolerance. To declare that a positional tolerance is to apply at MMC, a modifier is added to the positional tolerance in the feature control frame. Without the ⓜ modifier, the positional tolerance is held fixed, regardless of feature size (RFS); there is no bonus tolerance. In the RFS case, a standard fixed gage cannot be used to verify conformance.

Along with bonus tolerance, the concept of virtual condition has been previously introduced, and merits revisiting. For a position tolerance applied at MMC to a hole or other internal feature, that feature's virtual condition can be viewed as a theoretical boundary equivalent to a circle at true position with a diameter equal to the hole diameter at MMC minus the position tolerance. Since no element of the actual hole can be inside this theoretical boundary, the virtual condition of a hole feature is used to establish the diameter of a gage cylinder used to check the conformance of hole sizes and their locations against specifications.

In the case of a shaft or other external feature, the theoretical boundary (i.e., virtual condition) is equal to the shaft MMC plus the positional tolerance.

Hole virtual condition = hole MMC − positional tolerance

Shaft virtual condition = shaft MMC + positional tolerance

Concentricity describes a condition in which the median points of all diametrically opposed elements of a surface of revolution about an axis coincide with the axis of a datum feature. The median points must fall within a cylindrical tolerance zone associated with the datum axis feature. Since the location of a (theoretical) datum axis is so difficult to locate, it is easier to inspect for cylindricity or runout. Concentricity must always be applied to a feature of size, and it always applies regardless of feature size. Consequently concentricity cannot be modified to take advantage of bonus tolerances.

Symmetry shares many similarities with concentricity. It is the condition where the median points of all diametrically opposed elements of two or more feature surfaces are coincident with an axis or center plane of a datum feature. The median points must fall between two parallel planes, the parallel planes being symmetric about the datum feature axis or center plane. Like concentricity, symmetry must always be associated with a feature of size, it always applies regardless of feature size, and cannot be modified

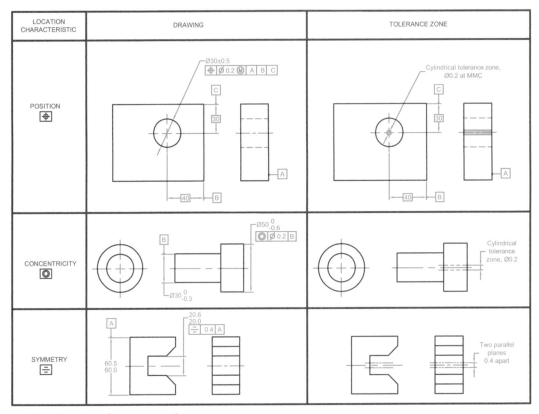

Figure B-31   Location tolerance examples

to MMC. In many situations positional tolerancing can provide a control for symmetry that is more flexible and easier to verify than the symmetry control.

Figure B-31 provides examples of the three location tolerance controls, position, concentricity, and symmetry.

☐ RUNOUT TOLERANCES

Runout results from placing a solid of revolution on a spindle (e.g., a lathe) and rotating the part about its central axis while measuring its surface deviation from perfect roundness with a dial indicator.[1] There are two kinds of runout, circular and total. In measuring circular runout, the position of the dial indicator is held in a fixed position normal to the controlled surface. Circular runout consequently measures deviation from perfect roundness at specific cross sections along the length of the surface of revolution. To measure total runout, on the other hand, the dial indicator is swept along the length of the part as the part is being rotated. The entire surface is thus controlled simultaneously, relative to a datum axis. Not surprisingly, total runout is a more demanding and expensive verification process than circular runout. Examples of both circular and total runout are provided in Figure B-32 on page 301.

[1]A dial indicator is an inspection gage used to measure positive and negative variation on features.

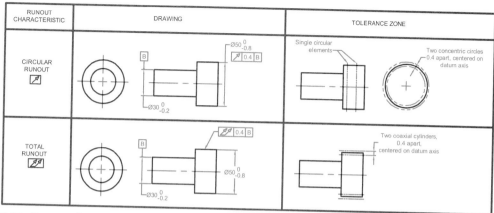

Figure B-32   Runout tolerance examples

ANSI PREFERRED ENGLISH LIMITS AND FITS

☐ ANSI RUNNING AND SLIDING FITS (RC)–ENGLISH UNITS

American National Standard Running and Sliding Fits (ANSI B4.1–1967, R1979)
Tolerance limits given in body of table are added or subtracted to basic size (as indicated by + or – sign) to obtain maximum and minimum sizes of mating parts.

Values shown below are in thousandths of an inch

Nominal Size Range, Inches (Over–To)	Class RC1 Clearance*	Class RC1 Hole H5	Class RC1 Shaft g4	Class RC2 Clearance*	Class RC2 Hole H6	Class RC2 Shaft g5	Class RC3 Clearance*	Class RC3 Hole H7	Class RC3 Shaft f6	Class RC4 Clearance*	Class RC4 Hole H8	Class RC4 Shaft f7
0–0.12	0.1 / 0.45	+0.2 / 0	−0.1 / −0.25	0.1 / 0.55	+0.25 / 0	−0.1 / −0.3	0.3 / 0.95	+0.4 / 0	−0.3 / −0.55	0.3 / 1.3	+0.6 / 0	−0.3 / −0.7
0.12–0.24	0.15 / 0.5	+0.2 / 0	−0.15 / −0.3	0.15 / 0.65	+0.3 / 0	−0.15 / −0.35	0.4 / 1.12	+0.5 / 0	−0.4 / −0.7	0.4 / 1.6	+0.7 / 0	−0.4 / −0.9
0.24–0.40	0.2 / 0.6	+0.25 / 0	−0.2 / −0.35	0.2 / 0.85	+0.4 / 0	−0.2 / −0.45	0.5 / 1.5	+0.6 / 0	−0.5 / −0.9	0.5 / 1.0	+0.9 / 0	−0.5 / −1.1
0.40–0.71	0.25 / 0.75	+0.3 / 0	−0.25 / −0.45	0.25 / 0.95	+0.4 / 0	−0.25 / −0.55	0.6 / 1.7	+0.7 / 0	−0.6 / −1.0	0.6 / 2.3	+1.0 / 0	−0.6 / −1.3
0.71–1.19	0.3 / 0.95	+0.4 / 0	−0.3 / −0.55	0.3 / 1.2	+0.5 / 0	−0.3 / −0.7	0.8 / 2.1	+0.8 / 0	−0.8 / −1.3	0.8 / 2.8	+1.2 / 0	−0.8 / −1.6
1.19–1.97	0.4 / 1.1	+0.4 / 0	−0.4 / −0.7	0.4 / 1.4	+0.6 / 0	−0.4 / −0.8	1.0 / 2.6	+1.0 / 0	−1.0 / −1.6	1.0 / 3.6	+1.6 / 0	−1.0 / −2.0
1.97–3.15	0.4 / 1.2	+0.5 / 0	−0.4 / −0.7	0.4 / 1.6	+0.7 / 0	−0.4 / −0.9	1.2 / 3.1	+1.2 / 0	−1.2 / −1.9	1.2 / 4.2	+1.8 / 0	−1.2 / −2.4
3.15–4.73	0.5 / 1.5	+0.6 / 0	−0.5 / −0.9	0.5 / 2.0	+0.9 / 0	−0.5 / −1.1	1.4 / 3.7	+1.4 / 0	−1.4 / −2.3	1.4 / 5.0	+2.2 / 0	−1.4 / −2.8
4.73–7.09	0.6 / 1.8	+0.7 / 0	−0.6 / −1.1	0.6 / 2.3	+1.0 / 0	−0.6 / −1.3	1.6 / 4.2	+1.6 / 0	−1.6 / −2.6	1.6 / 5.7	+2.5 / 0	−1.6 / −3.2
7.09–9.85	0.6 / 2.0	+0.8 / 0	−0.6 / −1.2	0.6 / 2.6	+1.2 / 0	−0.6 / −1.4	2.0 / 5.0	+1.8 / 0	−2.0 / −3.2	2.0 / 6.6	+2.8 / 0	−2.0 / −3.8
9.85–12.41	0.8 / 2.3	+0.9 / 0	−0.8 / −1.4	0.8 / 2.9	+1.2 / 0	−0.8 / −1.7	2.5 / 5.7	+2.0 / 0	−2.5 / −3.7	2.5 / 7.5	+3.0 / 0	−2.5 / −4.5
12.41–15.75	1.0 / 2.7	+1.0 / 0	−1.0 / −1.7	1.0 / 3.4	+1.4 / 0	−1.0 / −2.0	3.0 / 6.6	+2.2 / 0	−3.0 / −4.4	3.0 / 8.7	+3.5 / 0	−3.0 / −5.2
15.75–19.69	1.2 / 3.0	+1.0 / 0	−1.2 / −2.0	1.2 / 3.8	+1.6 / 0	−1.2 / −2.2	4.0 / 8.1	+2.5 / 0	−4.0 / −5.6	4.0 / 10.5	+4.0 / 0	−4.0 / −6.5

See footnotes at end of table.

Values shown below are in thousandths of an inch.

Nominal Size Range, Inches Over To	Class RC5 Clearance*	Class RC5 Hole H8	Class RC5 Shaft e7	Class RC6 Clearance*	Class RC6 Hole H9	Class RC6 Shaft e8	Class RC7 Clearance*	Class RC7 Hole H9	Class RC7 Shaft d8	Class RC8 Clearance*	Class RC8 Hole H10	Class RC8 Shaft c9	Class RC9 Clearance*	Class RC9 Hole H11	Class RC9 Shaft
0–0.12	0.6 1.6	+0.6 0	−0.6 −1.0	0.6 2.2	+1.0 0	−0.6 −1.2	1.0 2.6	+1.0 0	−1.0 −1.6	2.5 5.1	+1.6 0	−2.5 −3.5	4.0 8.1	+2.5 0	−4.0 −5.6
0.12–0.24	0.8 2.0	+0.7 0	−0.8 −1.3	0.8 2.7	+1.2 0	−0.8 −1.5	1.2 3.1	+1.2 0	−1.2 −1.9	2.8 5.8	+1.8 0	−2.8 −4.0	4.5 9.0	+3.0 0	−4.5 −6.0
0.24–0.40	1.0 2.5	+0.9 0	−1.0 −1.6	1.0 3.3	+1.4 0	−1.0 −1.9	1.6 3.9	+1.4 0	−1.6 −2.5	3.0 6.6	+2.2 0	−3.0 −4.4	5.0 10.7	+3.5 0	−5.0 −7.2
0.40–0.71	1.2 2.9	+1.0 0	−1.2 −1.9	1.2 3.8	+1.6 0	−1.2 −2.2	2.0 4.6	+1.6 0	−2.0 −3.0	3.5 7.9	+2.8 0	−3.5 −5.1	6.0 12.8	+4.0 0	−6.0 −8.8
0.71–1.19	1.6 3.6	+1.2 0	−1.6 −2.4	1.6 4.8	+2.0 0	−1.6 −2.8	2.5 5.7	+2.0 0	−2.5 −3.7	4.5 10.0	+3.5 0	−4.5 −6.5	7.0 15.5	+5.0 0	−7.0 −10.5
1.19–1.97	2.0 4.6	+1.6 0	−2.0 −3.0	2.0 6.1	+2.5 0	−2.0 −3.6	3.0 7.1	+2.5 0	−3.0 −4.6	5.0 11.5	+4.0 0	−5.0 −7.5	8.0 18.0	+6.0 0	−8.0 −12.0
1.97–3.15	2.5 5.5	+1.8 0	−2.5 −3.7	2.5 7.3	+3.0 0	−2.5 −4.3	4.0 8.8	+3.0 0	−4.0 −5.8	6.0 13.5	+4.5 0	−6.0 −9.0	9.0 20.5	+7.0 0	−9.0 −13.5
3.15–4.73	3.0 6.6	+2.2 0	−3.0 −4.4	3.0 8.7	+3.5 0	−3.0 −5.2	5.0 10.7	+3.5 0	−5.0 −7.2	7.0 15.5	+5.0 0	−7.0 −10.5	10.0 24.0	+9.0 0	−10.0 −15.0
4.73–7.09	3.5 7.6	+2.5 0	−3.5 −5.1	3.5 10.0	+4.0 0	−3.5 −6.0	6.0 12.5	+4.0 0	−6.0 −8.5	8.0 18.0	+6.0 0	−8.0 −12.0	12.0 28.0	+10.0 0	−12.0 −18.0
7.09–9.85	4.0 8.6	+2.8 0	−4.0 −5.8	4.0 11.3	+4.5 0	−4.0 −6.8	7.0 14.3	+4.5 0	−7.0 −9.8	10.0 21.5	+7.0 0	−10.0 −14.5	15.0 34.0	+12.0 0	−15.0 −22.0
9.85–12.41	5.0 10.0	+3.0 0	−5.0 −7.0	5.0 13.0	+5.0 0	−5.0 −8.0	8.0 16.0	+5.0 0	−8.0 −11.0	12.0 25.0	+8.0 0	−12.0 −17.0	18.0 38.0	+12.0 0	−18.0 −26.0
12.41–15.75	6.0 11.7	+3.5 0	−6.0 −8.2	6.0 15.5	+6.0 0	−6.0 −9.5	10.0 19.5	+6.0 0	−10.0 −13.5	14.0 29.0	+9.0 0	−14.0 −20.0	22.0 45.0	+14.0 0	−22.0 −31.0
15.75–19.69	8.0 14.5	+4.0 0	−8.0 −10.5	8.0 18.0	+6.0 0	−8.0 −12.0	12.0 22.0	+6.0 0	−12.0 −16.0	16.0 32.0	+10.0 0	−16.0 −22.0	25.0 51.0	+16.0 0	−25.0 −35.0

Data in bold type face are in accordance with American, British, Canadian (ABC) Agreements. Symbols H5, g4, etc. are hole and shaft designations in ABC system. Limits for sizes above 19.69 inches are also given in the ANSI Standard.

*Pairs of values shown represent minimum and maximum amounts of clearance resulting from application of standard tolerance limits.

Source: Reprinted courtesy of The American Society of Mechanical Engineers.

□ ANSI CLEARANCE LOCATION FITS (LC)–ENGLISH UNITS

American National Standard Clearance Locational Fits (ANSI B4.1-1967, R1979)

Tolerance limits given in body of table are added or subtracted to basic size (as indicated by + or – sign) to obtain maximum and minimum sizes of mating parts.

Values shown below are in thousandths of an inch

Nominal Size Range, Inches Over To	Class LC1 Clearance*	LC1 Hole H6	LC1 Shaft h5	Class LC2 Clearance*	LC2 Hole H7	LC2 Shaft h6	Class LC3 Clearance*	LC3 Hole H8	LC3 Shaft h7	Class LC4 Clearance*	LC4 Hole H10	LC4 Shaft h9	Class LC5 Clearance*	LC5 Hole H7	LC5 Shaft g6
0–0.12	0 / 0.45	+0.25 / 0	0 / −0.2	0 / 0.65	+0.4 / 0	0 / −0.25	0 / 1	+0.6 / 0	0 / −0.4	0 / 2.6	+1.6 / 0	0 / −1.0	0.1 / 0.75	+0.4 / 0	−0.1 / −0.35
0.12–0.24	0 / 0.5	+0.3 / 0	0 / −0.2	0 / 0.8	+0.5 / 0	0 / −0.3	0 / 1.2	+0.7 / 0	0 / −0.5	0 / 3.0	+1.8 / 0	0 / −1.2	0.15 / 0.95	+0.5 / 0	−0.15 / −0.45
0.24–0.40	0 / 0.65	+0.4 / 0	0 / −0.25	0 / 1.0	+0.6 / 0	0 / −0.4	0 / 1.5	+0.9 / 0	0 / −0.6	0 / 3.6	+2.2 / 0	0 / −1.4	0.2 / 1.2	+0.6 / 0	−0.2 / −0.6
0.40–0.71	0 / 0.7	+0.4 / 0	0 / −0.3	0 / 1.1	+0.7 / 0	0 / −0.4	0 / 1.7	+1.0 / 0	0 / −0.7	0 / 4.4	+2.8 / 0	0 / −1.6	0.25 / 1.35	+0.7 / 0	−0.25 / −0.65
0.71–1.19	0 / 0.9	+0.5 / 0	0 / −0.4	0 / 1.3	+0.8 / 0	0 / −0.5	0 / 2	+1.2 / 0	0 / −0.8	0 / 5.5	+3.5 / 0	0 / −2.0	0.3 / 1.6	+0.8 / 0	−0.3 / −0.8
1.19–1.97	0 / 1.0	+0.6 / 0	0 / −0.4	0 / 1.6	+1.0 / 0	0 / −0.6	0 / 2.6	+1.6 / 0	0 / −1	0 / 6.5	+4.0 / 0	0 / −2.5	0.4 / 2.0	+1.0 / 0	−0.4 / −1.0
1.97–3.15	0 / 1.2	+0.7 / 0	0 / −0.5	0 / 1.9	+1.2 / 0	0 / −0.7	0 / 3	+1.8 / 0	0 / −1.2	0 / 7.5	+4.5 / 0	0 / −3	0.4 / 2.3	+1.2 / 0	−0.4 / −1.1
3.15–4.73	0 / 1.5	+0.9 / 0	0 / −0.6	0 / 2.3	+1.4 / 0	0 / −0.9	0 / 3.6	+2.2 / 0	0 / −1.4	0 / 8.5	+5.0 / 0	0 / −3.5	0.5 / 2.8	+1.4 / 0	−0.5 / −1.4
4.73–7.09	0 / 1.7	+1.0 / 0	0 / −0.7	0 / 2.6	+1.6 / 0	0 / −1.0	0 / 4.1	+2.5 / 0	0 / −1.6	0 / 10.0	+6.0 / 0	0 / −4	0.6 / 3.2	+1.6 / 0	−0.6 / −1.6
7.09–9.85	0 / 2.0	+1.2 / 0	0 / −0.8	0 / 3.0	+1.8 / 0	0 / −1.2	0 / 4.6	+2.8 / 0	0 / −1.8	0 / 11.5	+7.0 / 0	0 / −4.5	0.6 / 3.6	+1.8 / 0	−0.6 / −1.8
9.85–12.41	0 / 2.1	+1.2 / 0	0 / −0.9	0 / 3.2	+2.0 / 0	0 / −1.2	0 / 5	+3.0 / 0	0 / −2.0	0 / 13.0	+8.0 / 0	0 / −5	0.7 / 3.9	+2.0 / 0	−0.7 / −1.9
12.41–15.75	0 / 2.4	+1.4 / 0	0 / −1.0	0 / 3.6	+2.2 / 0	0 / −1.4	0 / 5.7	+3.5 / 0	0 / −2.2	0 / 15.0	+9.0 / 0	0 / −6	0.7 / 4.3	+2.2 / 0	−0.7 / −2.1
15.75–19.69	0 / 2.6	+1.6 / 0	0 / −1.0	0 / 4.1	+2.5 / 0	0 / −1.6	0 / 6.5	+4 / 0	0 / −2.5	0 / 16.0	+10.0 / 0	0 / −6	0.8 / 4.9	+2.5 / 0	−0.8 / −2.4

See footnotes at end of table.

Values shown below are in thousandths of an inch

Nominal Size Range, Inches Over To	Class LC6 Clearance*	Class LC6 Std Tol. Hole H9	Class LC6 Std Tol. Shaft f8	Class LC7 Clearance*	Class LC7 Std Tol. Hole H10	Class LC7 Std Tol. Shaft e9	Class LC8 Clearance*	Class LC8 Std Tol. Hole H10	Class LC8 Std Tol. Shaft d9	Class LC9 Clearance*	Class LC9 Std Tol. Hole H11	Class LC9 Std Tol. Shaft c10	Class LC10 Clearance*	Class LC10 Std Tol. Hole H12	Class LC10 Std Tol. Shaft	Class LC11 Clearance*	Class LC11 Std Tol. Hole H13	Class LC11 Std Tol. Shaft
0–0.12	0.3 / 1.9	+1.0 / 0	-0.3 / -0.9	0.6 / 3.2	+1.6 / 0	-0.6 / -1.6	1.0 / 2.0	+1.6 / 0	-1.0 / -2.0	2.5 / 6.6	+2.5 / 0	-2.5 / -4.1	4 / 12	+4 / 0	-4 / -8	5 / 17	+6 / 0	-5 / -11
0.12–0.24	0.4 / 2.3	+1.2 / 0	-0.4 / -1.1	0.8 / 3.8	+1.8 / 0	-0.8 / -2.0	1.2 / 4.2	+1.8 / 0	-1.2 / -2.4	2.8 / 7.6	+3.0 / 0	-2.8 / -4.6	4.5 / 14.5	+5 / 0	-4.5 / -9.5	6 / 20	+7 / 0	-6 / -13
0.24–0.40	0.5 / 1.8	+1.4 / 0	-0.5 / -1.4	1.0 / 4.6	+2.3 / 0	-1.0 / -2.4	1.6 / 5.2	+2.2 / 0	-1.6 / -3.0	3.0 / 8.7	+3.5 / 0	-3.0 / -5.2	5 / 17	+6 / 0	-5 / -11	7 / 25	+9 / 0	-7 / -16
0.40–0.71	0.6 / 3.2	+1.6 / 0	-0.6 / -1.6	1.3 / 5.6	+2.8 / 0	-1.2 / -2.8	2.0 / 6.4	+2.8 / 0	-2.0 / -3.6	3.5 / 10.3	+4.0 / 0	-3.5 / -6.3	6 / 20	+7 / 0	-6 / -13	8 / 28	+10 / 0	-8 / -18
0.71–1.19	0.8 / 4.0	+2.0 / 0	-0.8 / -2.0	1.6 / 7.1	+3.5 / 0	-1.6 / -3.6	3.5 / 8.0	+3.5 / 0	-2.5 / -4.5	4.5 / 13.0	+5.0 / 0	-4.5 / -8.0	7 / 23	+8 / 0	-7 / -15	10 / 34	+12 / 0	-10 / -22
1.19–1.97	1.0 / 5.1	+2.5 / 0	-1.0 / -2.6	2.0 / 8.5	+4.0 / 0	-2.0 / -4.5	3.6 / 9.5	+4.0 / 0	-3.0 / -5.5	5.0 / 15.0	+6 / 0	-5.0 / -9.0	8 / 28	+10 / 0	-8 / -18	12 / 44	+16 / 0	-12 / -28
1.97–3.15	1.2 / 6.0	+3.0 / 0	-1.0 / -3.0	2.5 / 10.0	+4.5 / 0	-2.5 / -5.5	4.0 / 11.5	+4.5 / 0	-4.0 / -7.0	6.0 / 17.5	+7 / 0	-6.0 / -10.5	10 / 34	+12 / 0	-10 / -22	14 / 50	+18 / 0	-14 / -32
3.15–4.73	1.4 / 7.1	+3.5 / 0	-1.4 / -3.6	3.0 / 11.5	+5.0 / 0	-3.0 / -6.5	5.0 / 13.5	+5.0 / 0	-5.0 / -8.5	7 / 21	+9 / 0	-7 / -12	11 / 39	+14 / 0	-11 / -25	16 / 60	+22 / 0	-16 / -38
4.73–7.09	1.6 / 8.1	+4.0 / 0	-1.6 / -4.1	3.5 / 13.5	+6.0 / 0	-3.5 / -7.5	6 / 16	+6 / 0	-6 / -10	8 / 24	+10 / 0	-8 / -14	12 / 44	+16 / 0	-12 / -28	18 / 68	+25 / 0	-18 / -43
7.09–9.85	2.0 / 9.3	+4.5 / 0	-2.0 / -4.8	4.0 / 15.5	+7.0 / 0	-4.0 / -8.5	7 / 18.5	+7 / 0	-7 / -11.5	10 / 29	+12 / 0	-10 / -17	16 / 52	+18 / 0	-16 / -34	22 / 78	+28 / 0	-22 / -50
9.85–12.41	2.2 / 10.2	+5.0 / 0	-2.2 / -5.2	4.5 / 17.5	+8.0 / 0	-4.5 / -9.5	7 / 20	+8 / 0	-7 / -12	12 / 32	+12 / 0	-12 / -20	20 / 60	+20 / 0	-20 / -40	28 / 88	+30 / 0	-28 / -58
12.41–15.75	2.5 / 12.0	+6.0 / 0	-2.5 / -6.0	5.0 / 20.0	+9.0 / 0	-5 / -11	8 / 23	+9 / 0	-8 / -14	14 / 37	+14 / 0	-14 / -23	22 / 66	+22 / 0	-22 / -44	30 / 100	+35 / 0	-30 / -65
15.75–19.69	2.8 / 12.8	+6.0 / 0	-2.8 / -6.8	5.0 / 31.0	+10.0 / 0	-5 / -11	9 / 25	+10 / 0	-9 / -15	16 / 42	+16 / 0	-16 / -26	25 / 75	+25 / 0	-25 / -50	35 / 115	+40 / 0	-35 / -75

Data in bold type face are in accordance with American-British-Canadian (ABC) agreements. Symbols H6, H7, s6, etc. are hole and shaft designations in ABC system. Limits for sizes above 19.69 inches are not covered by ABC agreements but are given in the ANSI Standard.

*Pairs of values shown represent minimum and maximum amounts of interference resulting from application of standard tolerance limits.

Source: Reprinted courtesy of The American Society of Mechanical Engineers.

ANSI TRANSITION LOCATION FITS (LT)–ENGLISH UNITS

ANSI Standard Transition Location Fits (ANSI B4.1–1967, R1979)

Values shown below are in thousandths of an inch

Nominal Size Range, Inches (Over To)	Class LT1 Fit*	Class LT1 Hole H7	Class LT1 Shaft js6	Class LT2 Fit*	Class LT2 Hole H8	Class LT2 Shaft js7	Class LT3 Fit*	Class LT3 Hole H7	Class LT3 Shaft k6	Class LT4 Fit*	Class LT4 Hole H7	Class LT4 Shaft n6	Class LT5 Fit*	Class LT5 Hole H7	Class LT5 Shaft n6	Class LT6 Fit*	Class LT6 Hole H7	Class LT6 Shift n7
0–0.12	−0.12 / +0.52	+0.4 / 0	+0.12 / −0.12	−0.2 / +0.8	+0.6 / 0	+0.2 / −0.2										−0.65 / +0.15	+0.4 / 0	+0.65 / +0.25
0.12–0.24	−0.15 / +0.65	+0.5 / 0	+0.15 / −0.15	−0.25 / +0.95	+0.7 / 0	+0.25 / −0.25										−0.8 / +0.2	+0.5 / 0	+0.8 / +0.3
0.24–0.40	−0.2 / +0.8	+0.6 / 0	+0.2 / −0.2	−0.3 / +1.2	+0.9 / 0	+0.3 / −0.3	−0.5 / +0.5	+0.6 / 0	+0.5 / +0.1	−0.7 / +0.8	+0.9 / 0	+0.7 / +0.1	−0.8 / +0.2	+0.6 / 0	+0.8 / +0.4	−1.0 / +0.2	+0.6 / 0	+1.0 / +0.4
0.40–0.71	−0.2 / +0.9	+0.7 / 0	+0.2 / −0.2	−0.35 / +1.35	+1.0 / 0	+0.35 / −0.35	−0.5 / +0.6	+0.7 / 0	+0.5 / +0.1	−0.8 / +0.9	+1.0 / 0	+0.8 / +0.1	−0.9 / +0.2	+0.7 / 0	+0.9 / +0.5	−1.2 / +0.2	+0.7 / 0	+1.2 / +0.5
0.71–1.19	−0.25 / +1.05	+0.8 / 0	+0.25 / −0.25	−0.4 / +1.6	+1.2 / 0	+0.4 / −0.4	−0.6 / +0.7	+0.8 / 0	+0.6 / +0.1	−0.9 / +1.1	+1.2 / 0	+0.9 / +0.1	−1.1 / +0.2	+0.8 / 0	+1.1 / +0.6	−1.4 / +0.2	+0.8 / 0	+1.4 / +0.6
1.19–1.97	−0.3 / +1.3	+1.0 / 0	+0.3 / −0.3	−0.5 / +2.1	+1.6 / 0	+0.5 / −0.5	−0.7 / +0.9	+1.0 / 0	+0.7 / +0.1	−1.1 / +1.5	+1.6 / 0	+1.1 / +0.1	−1.3 / +0.3	+1.0 / 0	+1.3 / +0.7	−1.7 / +0.3	+1.0 / 0	+1.7 / +0.7
1.97–3.15	−0.3 / +1.5	+1.2 / 0	+0.3 / −0.3	−0.6 / +2.4	+1.8 / 0	+0.6 / −0.6	−0.8 / +1.1	+1.2 / 0	+0.8 / +0.1	−1.3 / +1.7	+1.8 / 0	+1.3 / +0.1	−1.5 / +0.4	+1.2 / 0	+1.5 / +0.8	−2.0 / +0.4	+1.2 / 0	+2.0 / +0.8
3.15–4.73	−0.4 / +1.8	+1.4 / 0	+0.4 / −0.4	−0.7 / +2.9	+2.2 / 0	+0.7 / −0.7	−1.0 / +1.3	+1.4 / 0	+1.0 / +0.1	−1.5 / +2.1	+2.2 / 0	+1.5 / +0.1	−1.9 / +0.4	+1.4 / 0	+1.9 / +1.0	−2.4 / +0.4	+1.4 / 0	+2.4 / +1.0
4.73–7.09	−0.5 / +2.1	+1.6 / 0	+0.5 / −0.5	−0.8 / +3.3	+2.5 / 0	+0.8 / −0.8	−1.1 / +1.5	+1.6 / 0	+1.1 / +0.1	−1.7 / +2.4	+2.5 / 0	+1.7 / +0.1	−2.2 / +0.4	+1.6 / 0	+2.2 / +1.2	−2.8 / +0.4	+1.6 / 0	+2.8 / +1.2
7.09–9.85	−0.6 / +2.4	+1.8 / 0	+0.6 / −0.6	−0.9 / +3.7	+2.8 / 0	+0.9 / −0.9	−1.4 / +1.6	+1.8 / 0	+1.4 / +0.2	−2.0 / +2.6	+2.8 / 0	+2.0 / +0.2	−2.6 / +0.4	+1.8 / 0	+2.6 / +1.4	−3.2 / +0.4	+1.8 / 0	+3.2 / +1.4
9.85–12.41	−0.6 / +2.6	+2.0 / 0	+0.6 / −0.6	−1.0 / +4.0	+3.0 / 0	+1.0 / −1.0	−1.4 / +1.8	+2.0 / 0	+1.4 / +0.2	−2.2 / +2.8	+3.0 / 0	+2.2 / +0.2	−2.6 / +0.6	+2.0 / 0	+2.6 / +1.4	−3.4 / +0.6	+2.0 / 0	+3.4 / +1.4
12.41–15.75	−0.7 / +2.9	+2.2 / 0	+0.7 / −0.7	−1.0 / +4.5	+3.5 / 0	+1.0 / −1.0	−1.6 / +2.0	+2.2 / 0	+1.6 / +0.2	−2.4 / +3.3	+3.5 / 0	+2.4 / +0.2	−3.0 / +0.6	+2.2 / 0	+3.0 / +1.6	−3.8 / +0.6	+2.2 / 0	+3.8 / +1.6
15.75–19.69	−0.8 / +3.3	+2.5 / 0	+0.8 / −0.8	−1.2 / +5.2	+4.0 / 0	+1.2 / −1.2	−1.8 / +2.3	+2.5 / 0	+1.8 / +0.2	−2.7 / +3.8	+4.0 / 0	+2.7 / +0.2	−3.4 / +0.7	+2.5 / 0	+3.4 / +1.8	−4.3 / +0.7	+2.5 / 0	+4.3 / +1.8

Data in bold type face are in accordance with American, British, Canadian (ABC) Agreements. Symbols H7, js6, etc. are hole and shaft designations in ABC system.

*Pairs of values shown represent maximum amount of interference (−) and maximum amount of clearance (+) resulting from application of standard tolerance limits.

Source: Reprinted courtesy of The American Society of Mechanical Engineers.

☐ ANSI INTERFERENCE LOCATIONAL FITS (LT)–ENGLISH UNITS

Nominal Size Range, Inches	Class LN1 Standard Limits			Class LN2 Standard Limits			Class LN3 Standard Limits		
	Limits of Interference	Hole H6	Shaft n5	Limits of Interference	Hole H7	Shaft p6	Limits of Interference	Hole H7	Shaft r6
Over To	Values shown below are in thousandths of an inch								
0–0.12	0 / 0.45	+0.25 / 0	+0.45 / +0.25	0 / 0.65	+0.4 / 0	+0.65 / +0.4	0.1 / 0.75	+0.4 / 0	+0.75 / +0.5
0.12–0.24	0 / 0.5	+0.3 / 0	+0.5 / +0.3	0 / 0.8	+0.5 / 0	+0.8 / +0.5	0.1 / 0.9	+0.5 / 0	+0.9 / +0.6
0.24–0.40	0 / 0.65	+0.4 / 0	+0.65 / +0.4	0 / 1.0	+0.6 / 0	+1.0 / +0.6	0.2 / 1.2	+0.6 / 0	+1.2 / +0.8
0.40–0.71	0 / 0.8	+0.4 / 0	+0.8 / +0.4	0 / 1.1	+0.7 / 0	+1.1 / +0.7	0.3 / 1.4	+0.7 / 0	+1.4 / +1.0
0.71–1.19	0 / 1.0	+0.5 / 0	+1.0 / +0.5	0 / 1.3	+0.8 / 0	+1.3 / +0.8	0.4 / 1.7	+0.8 / 0	+1.7 / +1.2
1.19–1.97	0 / 1.1	+0.6 / 0	+1.1 / +0.6	0 / 1.6	+1.0 / 0	+1.6 / +1.0	0.4 / 2.0	+1.0 / 0	+2.0 / +1.4
1.97–3.15	0.1 / 1.3	+0.7 / 0	+1.3 / +0.8	0.2 / 2.1	+1.2 / 0	+2.1 / +1.4	0.4 / 2.3	+1.2 / 0	+2.3 / +1.6
3.15–4.73	0.1 / 1.6	+0.9 / 0	+1.6 / +1.0	0.2 / 2.5	+1.4 / 0	+2.5 / +1.6	0.6 / 2.9	+1.4 / 0	+2.9 / +2.0
4.73–7.09	0.2 / 1.9	+1.0 / 0	+1.9 / +1.2	0.2 / 2.8	+1.6 / 0	+2.8 / +1.8	0.9 / 3.5	+1.6 / 0	+3.5 / +2.5
7.09–9.85	0.1 / 2.2	+1.2 / 0	+2.2 / +1.4	0.2 / 3.2	+1.8 / 0	+3.2 / +2.0	1.2 / 4.2	+1.8 / 0	+4.2 / +3.0
9.85–12.41	0.2 / 2.3	+1.2 / 0	+2.3 / +1.4	0.2 / 3.4	+2.0 / 0	+3.4 / +2.2	1.5 / 4.7	+2.0 / 0	+4.7 / +3.5
12.41–15.75	0.2 / 2.6	+1.4 / 0	+2.6 / +1.6	0.3 / 3.9	+2.2 / 0	+3.9 / +2.5	2.3 / 5.9	+2.2 / 0	+5.9 / +4.5
15.75–19.69	0.2 / 1.8	+1.6 / 0	+2.8 / +1.8	0.3 / 4.4	+2.5 / 0	+4.4 / +2.8	2.5 / 6.6	+2.5 / 0	+6.6 / +5.0

All data in this table are in accordance with American-British-Canadian (ABC) agreements.
Limits for sizes above 19.69 inches are not covered by ABC agreements but are given in the ANSI Standard.
Symbols H7, p6, etc. are hole and shaft designations in ABC system.
*Pairs of values shown represent minimum and maximum amounts of interference resulting from application of standard tolerance limits.
Source: Reprinted courtesy of The American Society of Mechanical Engineers.

☐ **ANSI FORCE AND SHRINK FITS (FN)–ENGLISH UNITS**

ANSI Standard Force and Shrink Fits (ANSI B4.1–1967, R1979)

Values shown below are in thousandths of an inch

Nominal Size Range, Inches Over To	Class FN1 Inter-ference*	Class FN1 Hole H6	Class FN1 Shaft	Class FN2 Inter-ference*	Class FN2 Hole H7	Class FN2 Shaft s6	Class FN3 Inter-ference*	Class FN3 Hole H7	Class FN3 Shaft 16	Class FN4 Inter-ference*	Class FN4 Hole H7	Class FN4 Shaft u6	Class FN5 Inter-ference*	Class FN5 Hole H8	Class FN5 Shaft x7
0–0.12	0.05 / 0.5	+0.25 / 0	+0.5 / +0.3	0.2 / 0.85	+0.4 / 0	+0.85 / +0.6				0.3 / 0.95	+0.4 / 0	+0.95 / +0.7	0.3 / 1.3	+0.6 / 0	+1.3 / +0.9
0.12–0.24	0.1 / 0.6	+0.3 / 0	+0.6 / +0.4	0.2 / 1.0	+0.5 / 0	+1.0 / +0.7				0.4 / 1.2	+0.5 / 0	+1.2 / +0.9	0.5 / 1.7	+0.7 / 0	+1.7 / +1.2
0.24–0.40	0.1 / 0.75	+0.4 / 0	+0.75 / +0.5	0.4 / 1.4	+0.6 / 0	+1.4 / +1.0				0.6 / 1.6	+0.6 / 0	+1.6 / +1.2	0.5 / 2.0	+0.9 / 0	+2.0 / +1.4
0.40–0.56	0.1 / 0.8	+0.4 / 0	+0.8 / +0.5	0.5 / 1.6	+0.7 / 0	+1.6 / +1.2				0.7 / 1.8	+0.7 / 0	+1.8 / +1.4	0.6 / 2.3	+1.0 / 0	+2.3 / +1.6
0.56–0.71	0.2 / 0.9	+0.4 / 0	+0.9 / +0.6	0.5 / 1.6	+0.7 / 0	+1.6 / +1.2				0.7 / 1.8	+0.7 / 0	+1.8 / +1.4	0.8 / 2.5	+1.0 / 0	+2.5 / +1.8
0.71–0.95	0.2 / 1.1	+0.5 / 0	+1.1 / +0.7	0.6 / 1.9	+0.8 / 0	+1.9 / +1.4				0.8 / 2.1	+0.8 / 0	+2.1 / +1.6	1.0 / 3.0	+1.2 / 0	+3.0 / +2.2
0.95–1.19	0.3 / 1.2	+0.5 / 0	+1.2 / +0.8	0.6 / 1.9	+0.8 / 0	+1.9 / +1.4	0.8 / 2.1	+0.8 / 0	+2.1 / +1.6	1.0 / 2.3	+0.8 / 0	+2.3 / +1.8	1.3 / 3.3	+1.2 / 0	+3.3 / +2.5
1.19–1.58	0.3 / 1.3	+0.6 / 0	+1.3 / +0.9	0.8 / 2.4	+1.0 / 0	+2.4 / +1.8	1.0 / 2.6	+1.0 / 0	+2.6 / +2.0	1.5 / 3.1	+1.0 / 0	+3.1 / +2.5	1.4 / 4.0	+1.6 / 0	+4.0 / +3.0
1.58–1.97	0.4 / 1.4	+0.6 / 0	+1.4 / +1.0	0.8 / 2.4	+1.0 / 0	+2.4 / +1.8	1.2 / 2.8	+1.0 / 0	+2.8 / +2.2	1.8 / 3.4	+1.0 / 0	+3.4 / +2.8	2.4 / 5.0	+1.6 / 0	+5.0 / +4.0
1.97–2.56	0.6 / 1.8	+0.7 / 0	+1.8 / +1.3	0.8 / 2.7	+1.2 / 0	+2.7 / +2.0	1.3 / 3.2	+1.2 / 0	+3.2 / +2.5	2.3 / 4.2	+1.2 / 0	+4.2 / +3.5	3.2 / 6.2	+1.8 / 0	+6.2 / +5.0
2.56–3.15	0.7 / 1.9	+0.7 / 0	+1.9 / +1.4	1.0 / 2.9	+1.2 / 0	+2.9 / +2.2	1.8 / 3.7	+1.2 / 0	+3.7 / +3.0	2.8 / 4.7	+1.2 / 0	+4.7 / +4.0	4.2 / 7.2	+1.8 / 0	+7.2 / +6.0
3.15–3.94	0.9 / 2.4	+0.9 / 0	+2.4 / +1.8	1.4 / 3.7	+1.4 / 0	+3.7 / +2.8	2.1 / 4.4	+1.4 / 0	+4.4 / +3.5	3.6 / 5.9	+1.4 / 0	+5.9 / +5.0	4.8 / 8.4	+2.2 / 0	+8.4 / +7.0
3.94–4.73	1.1 / 2.6	+0.9 / 0	+2.6 / +2.0	1.6 / 3.9	+1.4 / 0	+3.9 / +3.0	2.6 / 4.9	+1.4 / 0	+4.9 / +4.0	4.6 / 6.9	+1.4 / 0	+6.9 / +6.0	5.8 / 9.4	+2.2 / 0	+9.4 / +8.0

See footnotes at end of table.

Values shown below are in thousandths of an inch.

Nominal Size Range, Inches Over To	Class FN Interference*	Class FN Standard Tolerance Limits Hole H6	Class FN Standard Tolerance Limits Shaft	Class FN Interference*	Class FN Standard Tolerance Limits Hole H7	Class FN Standard Tolerance Limits Shaft s6	Class FN Interference*	Class FN Standard Tolerance Limits Hole H7	Class FN Standard Tolerance Limits Shaft 16	Class FN Interference*	Class FN Standard Tolerance Limits Hole H7	Class FN Standard Tolerance Limits Shaft u6	Class FN Interference*	Class FN Standard Tolerance Limits Hole H8	Class FN Standard Tolerance Limits Shaft x7
4.73–5.52	1.2 / 2.9	+1.0 / 0	+2.9 / +2.2	1.9 / 4.5	+1.6 / 0	+4.5 / +3.5	3.4 / 6.0	+1.6 / 0	+6.0 / +5.0	5.4 / 8.0	+1.6 / 0	+8.0 / +7.0	7.5 / 11.6	+2.5 / 0	+11.6 / +10.0
5.52–6.30	1.5 / 3.2	+1.0 / 0	+3.2 / +2.5	2.4 / 5.0	+1.6 / 0	+5.0 / +4.0	3.4 / 6.0	+1.6 / 0	+6.0 / +5.0	5.4 / 8.0	+1.6 / 0	+8.0 / +7.0	9.5 / 13.6	+2.5 / 0	+13.6 / +12.0
6.30–7.09	1.8 / 3.5	+1.0 / 0	+3.5 / +2.8	2.9 / 5.5	+1.6 / 0	+5.5 / +4.5	4.4 / 7.0	+1.6 / 0	+7.0 / +6.0	6.4 / 9.0	+1.6 / 0	+9.0 / +8.0	9.5 / 13.6	+2.5 / 0	+13.6 / +12.0
7.09–7.88	1.8 / 3.8	+1.2 / 0	+3.8 / +3.0	3.2 / 6.2	+1.8 / 0	+6.2 / +5.0	5.2 / 8.2	+1.8 / 0	+8.2 / +7.0	7.2 / 10.2	+1.8 / 0	+10.2 / +9.0	11.2 / 15.8	+2.8 / 0	+15.8 / +14.0
7.88–8.86	2.3 / 4.3	+1.2 / 0	+4.3 / +3.5	3.2 / 6.2	+1.8 / 0	+6.2 / +5.0	5.2 / 6.2	+1.8 / 0	+8.2 / +7.0	8.2 / 11.2	+1.8 / 0	+11.2 / +10.0	13.2 / 17.8	+2.8 / 0	+17.8 / +16.0
8.86–9.85	2.3 / 4.3	+1.2 / 0	+4.3 / +3.5	4.2 / 7.2	+1.8 / 0	+7.2 / +6.0	6.2 / 9.2	+1.8 / 0	+9.2 / +8.0	10.2 / 13.2	+1.8 / 0	+13.2 / +12.0	13.2 / 17.8	+2.8 / 0	+17.8 / +16.0
9.85–11.03	2.8 / 4.9	+1.2 / 0	+4.9 / +4.0	4.0 / 7.2	+2.0 / 0	+7.2 / +6.0	7.0 / 10.2	+2.0 / 0	+10.2 / +9.0	10.0 / 13.2	+2.0 / 0	+13.2 / +12.0	15.0 / 20.0	+3.0 / 0	+20.0 / +18.0
11.03–12.41	2.8 / 4.9	+1.2 / 0	+4.9 / +4.0	5.0 / 8.2	+2.0 / 0	+8.2 / +7.2	7.0 / 10.2	+2.0 / 0	+10.2 / +9.0	12.0 / 15.2	+2.0 / 0	+15.2 / +14.0	17.0 / 22.0	+3.0 / 0	+22.0 / +20.0
12.41–13.98	3.1 / 5.5	+1.4 / 0	+5.5 / +4.5	5.8 / 9.4	+2.2 / 0	+9.4 / +8.0	7.8 / 11.4	+2.2 / 0	+11.4 / +10.0	13.8 / 17.4	+2.2 / 0	+17.4 / +16.0	18.5 / 24.2	+3.5 / 0	+24.2 / +22.0
13.98–15.75	3.6 / 6.1	+1.4 / 0	+6.1 / +5.0	5.8 / 9.4	+2.2 / 0	+9.4 / +8.0	9.8 / 13.4	+2.2 / 0	+13.4 / +12.0	15.8 / 19.4	+2.2 / 0	+19.4 / +18.0	21.5 / 27.2	+3.5 / 0	+27.2 / +25.0
15.75–17.72	4.4 / 7.0	+1.6 / 0	+7.0 / +6.0	6.5 / 10.6	+2.5 / 0	+10.6 / +9.0	9.5 / 13.6	+2.5 / 0	+13.6 / +12.0	17.5 / 21.6	+2.5 / 0	+21.6 / +20.0	24.0 / 30.5	+4.0 / 0	+30.5 / +28.0
17.72–19.69	4.4 / 7.0	+1.6 / 0	+7.0 / +6.0	7.5 / 11.6	+2.5 / 0	+11.6 / +10.0	11.5 / 15.6	+2.5 / 0	+15.6 / +14.0	19.5 / 23.6	+2.5 / 0	+23.6 / +22.0	26.0 / 32.5	+4.0 / 0	+32.5 / +30.0

Data in bold type face are in accordance with American-British-Canadian (ABC) agreements. Symbols H6, H7, s6, etc. are hole and shaft designations in ABC system. Limits for sizes above 19-69 inches are not covered by ABC agreements but are given in the ANSI standard.

*Pairs of values shown represent minimum and maximum amounts of interference resulting from application of standard tolerance limits.

Source: Reprinted courtesy of The American Society of Mechanical Engineers.

D ANSI PREFERRED METRIC LIMITS AND FITS

□ ANSI PREFERRED HOLE BASIS METRIC CLEARANCE FITS—METRIC UNITS

American National Standard Preferred Hole Basis Metric Clearance Fits (ANSI B4.2–1978, R1984)

Basic Size		Loose Running			Free Running			Close Running			Sliding			Locational Clearance		
		Hole H11	Shaft c11	Fit†	Hole H9	Shaft d9	Fit†	Hole H8	Shaft f7	Fit†	Hole H7	Shaft g6	Fit†	Hole H7	Shaft h6	Fit†
1	Max	1.060	0.940	0.180	1.025	0.980	0.070	1.014	0.994	0.030	1.010	0.998	0.018	1.010	1.000	0.016
	Min	1.000	0.880	0.060	1.000	0.955	0.020	1.000	0.984	0.006	1.000	0.992	0.002	1.000	0.994	0.000
1.2	Max	1.260	1.140	0.180	1.225	1.180	0.070	1.214	1.194	0.030	1.210	1.198	0.018	1.210	1.200	0.016
	Min	1.200	1.080	0.060	1.200	1.155	0.020	1.200	1.184	0.006	1.200	1.192	0.002	1.200	1.194	0.000
1.6	Max	1.660	1.540	0.180	1.625	1.580	0.070	1.614	1.594	0.030	1.610	1.598	0.018	1.610	1.600	0.016
	Min	1.600	1.480	0.060	1.600	1.555	0.020	1.600	1.584	0.006	1.600	1.592	0.002	1.600	1.594	0.000
2	Max	2.060	1.940	0.180	2.025	1.980	0.070	2.014	1.994	0.030	2.010	1.992	0.018	2.010	2.000	0.016
	Min	2.000	1.880	0.060	2.000	1.955	0.020	2.000	1.984	0.006	2.000	1.992	0.002	2.000	1.994	0.000
2.5	Max	2.560	2.440	0.180	2.525	2.480	0.070	2.514	2.494	0.030	2.510	2.498	0.018	2.510	2.500	0.016
	Min	2.500	2.380	0.060	2.500	2.455	0.020	2.500	2.484	0.006	2.500	2.492	0.002	2.500	2.494	0.000
3	Max	3.060	2.940	0.180	3.025	2.980	0.070	3.014	2.994	0.030	3.010	2.993	0.018	3.010	3.000	0.016
	Min	3.000	2.880	0.060	3.000	2.955	0.020	3.000	2.984	0.006	3.000	2.991	0.002	3.000	2.994	0.000
4	Max	4.075	3.930	0.220	4.030	3.970	0.090	4.018	3.990	0.040	4.012	3.996	0.024	4.012	4.000	0.020
	Min	4.000	3.855	0.070	4.000	3.940	0.030	4.000	3.978	0.010	4.000	3.988	0.004	4.000	3.992	0.000
5	Max	5.075	4.930	0.220	5.030	4.970	0.090	5.018	4.990	0.040	5.012	4.996	0.024	5.012	5.000	0.020
	Min	5.000	4.855	0.070	5.000	4.940	0.030	5.000	4.978	0.010	5.000	4.988	0.004	5.000	4.991	0.000
6	Max	6.075	5.930	0.220	6.030	5.970	0.090	6.018	5.990	0.040	6.012	5.996	0.024	6.012	6.000	0.020
	Min	6.000	5.855	0.070	6.000	5.940	0.030	6.000	5.978	0.010	6.000	5.988	0.004	6.000	5.992	0.000
8	Max	8.090	7.920	0.260	8.036	7.960	0.112	8.022	7.987	0.050	8.015	7.995	0.029	8.015	8.000	0.024
	Min	8.000	7.830	0.080	8.000	7.924	0.040	8.000	7.972	0.013	8.000	7.986	0.005	8.000	7.991	0.000
10	Max	10.090	9.920	0.260	10.036	9.960	0.112	10.022	9.987	0.050	10.015	9.995	0.029	10.015	10.000	0.024
	Min	10.000	9.830	0.080	10.000	9.924	0.040	10.000	9.972	0.013	10.000	9.986	0.005	10.000	9.991	0.000
12	Max	12.110	11.905	0.315	12.043	11.950	0.136	12.027	11.984	0.061	12.018	11.994	0.035	12.018	12.000	0.029
	Min	12.000	11.795	0.095	12.000	11.907	0.050	12.000	11.966	0.016	12.000	11.983	0.006	12.000	11.989	0.000
16	Max	16.110	15.905	0.315	16.043	15.950	0.136	16.027	15.984	0.061	16.018	15.994	0.035	16.018	16.000	0.029
	Min	16.000	15.795	0.095	16.000	15.907	0.050	16.000	15.966	0.016	16.000	15.983	0.006	16.000	15.989	0.000
20	Max	20.130	19.890	0.370	20.052	19.935	0.169	20.033	19.980	0.074	20.021	19.993	0.042	20.021	20.000	0.034
	Min	20.000	19.760	0.110	20.000	19.885	0.065	20.000	19.959	0.020	20.000	19.980	0.007	20.000	19.987	0.000

Basic Size		Loose Running			Free Running			Close Running			Sliding			Locational Clearance		
		Hole H11	Shaft c11	Fit[†]	Hole H9	Shaft d9	Fit[†]	Hole H8	Shaft f7	Fit[†]	Hole H7	Shaft g6	Fit[†]	Hole H7	Shaft h6	Fit[†]
25	Max	25.130	24.890	0.370	25.052	24.935	0.169	25.033	24.980	0.074	25.021	24.993	0.041	25.001	15.000	0.034
	Min	25.000	24.760	0.110	25.000	24.383	0.065	25.000	24.959	0.010	25.000	24.980	0.007	25.000	24.987	0.000
30	Max	30.130	29.890	0.370	30.052	29.935	0.169	30.033	29.980	0.074	30.021	29.993	0.041	30.031	30.000	0.034
	Min	30.000	29.760	0.110	30.000	19.883	0.065	30.000	29.959	0.020	30.000	29.980	0.007	30.000	29.987	0.000
40	Max	40.160	39.880	0.440	40.062	39.920	0.204	40.039	39.975	0.089	40.025	39.991	0.050	40.025	40.000	0.041
	Min	40.000	39.720	0.120	40.000	39.858	0.080	40.000	39.950	0.025	40.000	39.975	0.009	40.000	39.984	0.000
50	Max	50.160	49.870	0.450	50.062	49.920	0.204	50.039	49.975	0.089	50.025	49.991	0.050	50.025	50.000	0.041
	Min	50.000	49.710	0.130	50.000	49.858	0.080	50.000	49.950	0.025	50.000	49.975	0.009	50.000	49.984	0.000
60	Max	60.190	59.860	0.520	60.074	59.900	0.248	60.046	59.970	0.106	60.030	59.990	0.059	60.030	60.000	0.049
	Min	60.000	39.670	0.140	60.000	59.826	0.100	60.000	59.940	0.030	60.000	59.971	0.010	60.000	59.981	0.000
80	Max	80.190	79.850	0.530	80.074	79.900	0.248	80.046	19.970	0.106	80.090	78.990	0.059	80.030	80.000	0.049
	Min	80.000	79.660	0.150	80.000	79.826	0.100	80.000	79.940	0.030	80.000	79.971	0.010	80.000	79.981	0.000
100	Max	100.220	99.830	0.610	100.087	99.880	0.294	100.054	99.954	0.125	100.035	99.988	0.069	100.035	100.000	0.057
	Min	100.000	99.610	0.170	100.000	99.793	0.120	100.000	99.929	0.036	100.000	99.966	0.012	100.000	99.978	0.000
120	Max	210.230	119.820	0.620	120.087	119.880	0.294	120.054	119.964	0.125	120.033	119.988	0.069	120.035	120.000	0.057
	Min	110.000	119.600	0.180	120.000	119.793	0.120	120.000	119.929	0.036	120.000	119.966	0.012	120.000	119.978	0.000
160	Max	160.250	159.790	0.710	160.100	159.855	0.345	160.053	159.957	0.146	160.040	159.986	0.079	160.040	160.000	0.065
	Min	160.000	159.540	0.210	160.000	159.755	0.145	160.000	159.917	0.043	160.000	159.961	0.014	160.000	159.975	0.000
200	Max	200.290	199.760	0.820	200.115	119.830	0.400	200.072	199.950	0.168	200.046	199.985	0.090	200.046	200.000	0.071
	Min	200.000	199.470	0.240	200.000	199.715	0.170	200.000	199.904	0.050	200.000	199.956	0.015	200.000	199.971	0.000
250	Max	250.290	249.720	0.860	250.115	249.830	0.400	250.072	249.950	0.168	250.046	249.985	0.090	250.046	250.000	0.075
	Min	250.000	249.430	0.230	250.000	249.115	0.170	250.000	249.904	0.050	250.000	249.956	0.015	250.000	249.971	0.000
300	Max	300.320	299.670	0.970	300.130	299.810	0.450	300.081	299.944	0.189	300.052	299.983	0.101	300.052	300.000	0.084
	Min	300.000	299.350	0.330	300.000	299.680	0.190	300.000	299.892	0.056	300.000	299.951	0.017	300.000	299.968	0.000
400	Max	400.360	399.600	1.120	400.140	399.790	0.490	400.059	399.938	0.208	400.057	399.982	0.111	400.051	400.000	0.093
	Min	400.000	399.240	0.400	400.000	399.650	0.210	400.000	399.881	0.063	400.000	399.946	0.018	400.000	399.964	0.000
500	Max	500.400	499.520	1.280	500.155	499.770	0.540	500.097	499.932	0.228	500.063	499.80	0.123	500.063	500.000	0.103
	Min	500.000	499.120	0.480	500.000	499.615	0.230	500.000	499.869	0.068	500.000	499.940	0.020	500.000	499.960	0.000

All dimensions are in millimeters.

Preferred fits for other sizes can be calculated from data given in ANSI B4.2–1978 (R1984).

[†]All fits shown in this table have clearance.

Source: Reprinted courtesy of The American Society of Mechanical Engineers.

ANSI PREFERRED HOLE BASIS TRANSITION AND INTERFERENCE FITS—METRIC UNITS

American National Standard Preferred Hole Basis Metric Transition and Interference Fits (ANSI B4.2–1978, R1984)

Basic Size		Locational Transition Hole H7	Shaft k6	Fit†	Locational Transition Hole H7	Shaft n6	Fit†	Locational Interference Hole H7	Shaft p6	Fit†	Medium Drive Hole H7	Shaft s6	Fit†	Force Hole H7	Shaft u6	Fit†
1	Max	1.010	1.006	+0.010	1.010	1.010	+0.006	1.010	1.012	+0.004	1.010	1.020	-0.004	1.010	1.024	-0.008
	Min	1.000	1.000	-0.006	1.000	1.004	-0.010	1.000	1.006	-0.012	1.000	1.014	-0.020	1.000	1.018	-0.024
1.2	Max	1.210	1.206	+0.010	1.210	1.210	+0.006	1.210	1.212	+0.004	1.210	1.220	-0.004	1.210	1.224	-0.008
	Min	1.200	1.200	-0.006	1.200	1.204	-0.010	1.200	1.206	-0.012	1.200	1.214	-0.020	1.200	1.218	-0.024
1.6	Max	1.610	1.606	+0.010	1.610	1.610	+0.006	1.610	1.612	+0.004	1.610	1.620	-0.004	1.610	1.624	-0.008
	Min	1.600	1.600	-0.006	1.600	1.604	-0.010	1.600	1.605	-0.012	1.600	1.614	-0.020	1.600	1.618	-0.024
2	Max	2.010	2.006	+0.010	2.010	2.010	+0.006	2.010	2.012	+0.004	2.010	2.020	-0.004	2.010	2.024	-0.008
	Min	2.000	2.000	-0.006	2.000	2.004	-0.010	2.000	2.006	-0.012	2.000	2.014	-0.020	2.000	2.018	-0.024
2.5	Max	2.510	2.506	+0.010	2.510	2.510	+0.006	2.510	2.512	+0.004	2.510	2.520	-0.004	2.510	2.524	-0.008
	Min	2.500	2.500	-0.006	2.500	2.504	-0.010	2.500	2.506	-0.012	2.500	2.514	-0.020	2.500	2.518	-0.024
3	Max	3.010	3.006	+0.010	3.010	3.010	+0.006	3.010	3.012	+0.004	3.010	3.020	-0.004	3.010	3.024	-0.008
	Min	3.000	3.000	-0.006	3.000	3.004	-0.010	3.000	3.006	-0.012	3.000	3.014	-0.020	3.000	3.018	-0.024
4	Max	4.012	4.009	+0.011	4.012	4.016	+0.004	4.012	4.020	0.000	4.012	4.027	-0.007	4.012	4.031	-0.011
	Min	4.000	4.001	-0.009	4.000	4.008	-0.016	4.000	4.012	-0.020	4.000	4.019	-0.027	4.000	4.023	-0.031
5	Max	5.012	5.009	+0.011	5.012	5.016	+0.004	5.012	5.020	0.000	5.012	5.027	-0.007	5.012	5.031	-0.011
	Min	5.000	5.001	-0.009	5.000	5.008	-0.016	5.000	5.012	-0.020	5.000	5.019	-0.027	5.000	5.023	-0.031
6	Max	6.012	6.009	+0.011	6.012	6.016	+0.004	6.012	6.020	0.000	6.012	6.027	-0.007	6.012	6.031	-0.011
	Min	6.000	6.001	-0.009	6.000	6.008	-0.016	6.000	6.012	-0.020	6.000	6.019	-0.027	6.000	6.023	-0.031
8	Max	8.015	8.010	+0.014	8.015	8.019	+0.005	8.015	8.024	0.000	8.015	8.032	-0.008	8.015	8.037	-0.013
	Min	8.000	8.001	-0.010	8.000	8.010	-0.019	8.000	8.015	-0.024	8.000	8.023	-0.032	8.000	8.028	-0.037
10	Max	10.015	10.010	+0.014	10.015	10.019	+0.005	10.015	10.024	0.000	10.015	10.032	-0.008	10.015	10.037	-0.013
	Min	10.000	10.001	-0.010	10.000	10.010	-0.019	10.000	10.015	-0.024	10.000	10.023	-0.032	10.000	10.028	-0.037
12	Max	12.018	12.012	+0.017	12.018	12.023	+0.006	12.018	12.029	0.000	12.018	12.039	-0.010	12.018	12.044	-0.015
	Min	12.000	12.001	-0.012	12.000	12.012	-0.023	12.000	12.018	-0.029	12.000	12.028	-0.039	12.000	12.033	-0.044
16	Max	16.018	16.012	+0.017	16.018	16.023	+0.006	16.018	16.029	0.000	16.018	16.039	-0.010	16.018	16.044	-0.015
	Min	16.000	16.001	-0.012	16.000	16.012	-0.023	16.000	16.018	-0.029	16.000	16.028	-0.039	16.000	16.033	-0.044
20	Max	20.021	20.015	+0.019	20.021	20.018	+0.005	20.021	20.035	-0.001	20.021	20.048	-0.014	20.021	20.054	-0.020
	Min	20.000	20.002	-0.015	20.000	20.015	-0.028	20.000	20.012	-0.035	20.000	20.035	-0.048	20.000	20.041	-0.054

Basic Size		Locational Transition			Locational Transition			Locational Interference			Medium Drive			Force		
		Hole H7	Shaft k6	Fit[†]	Hole H7	Shaft n6	Fit[†]	Hole H7	Shaft p6	Fit[†]	Hole H7	Shaft s6	Fit[†]	Hole H7	Shaft u6	Fit[†]
25	Max	15.021	25.015	+0.019	25.021	25.028	+0.006	25.021	25.035	−0.001	25.021	25.048	−0.014	25.021	25.061	−0.027
	Min	25.000	25.002	−0.015	25.000	25.015	−0.028	25.000	25.022	−0.035	25.000	25.035	−0.048	25.000	25.048	−0.061
30	Max	30.021	30.015	+0.019	30.021	30.028	+0.006	30.021	30.035	−0.001	30.021	30.048	−0.014	30.021	30.051	−0.027
	Min	30.000	30.002	−0.015	30.000	30.015	−0.028	30.000	30.022	−0.035	30.000	30.035	−0.048	30.000	30.048	−0.061
40	Max	40.025	40.018	+0.023	40.025	40.033	+0.008	40.025	40.042	−0.001	40.025	40.059	−0.018	40.025	40.076	−0.035
	Min	40.000	40.002	−0.018	40.000	40.017	−0.033	40.000	40.026	−0.042	40.000	40.043	−0.059	40.000	40.060	−0.076
50	Max	50.025	50.018	+0.023	50.025	50.033	+0.008	50.025	50.042	−0.002	50.025	50.059	−0.018	50.025	50.086	−0.045
	Min	50.000	50.002	−0.018	50.000	50.017	−0.033	50.000	50.026	−0.042	50.000	50.043	−0.059	50.000	50.070	−0.086
60	Max	60.030	60.021	+0.028	60.030	60.039	+0.010	60.030	60.051	−0.002	60.030	60.072	−0.023	60.030	60.106	−0.057
	Min	60.000	60.002	−0.021	60.000	60.020	−0.039	60.000	60.032	−0.052	60.000	60.053	−0.072	60.000	60.087	−0.106
80	Max	80.030	80.021	+0.028	80.030	80.039	+0.010	80.030	80.051	−0.002	80.030	80.078	−0.029	80.030	80.121	−0.072
	Min	80.000	80.002	−0.021	80.000	80.020	−0.039	80.000	80.032	−0.051	80.000	80.059	−0.078	80.000	80.102	−0.121
100	Max	100.035	100.025	+0.032	100.035	100.045	+0.012	100.035	100.059	−0.002	100.035	100.093	−0.036	100.035	100.146	−0.089
	Min	100.000	100.003	−0.025	100.000	100.023	−0.045	100.000	100.037	−0.059	100.000	100.071	−0.093	100.000	100.124	−0.146
120	Max	120.035	120.025	+0.032	120.035	120.045	+0.012	120.035	120.059	−0.002	120.035	120.101	−0.044	120.035	120.166	−0.109
	Min	120.000	120.003	−0.025	120.000	120.023	−0.045	120.000	120.037	−0.059	120.000	120.079	−0.101	120.000	120.144	−0.166
160	Max	160.040	160.028	+0.037	160.040	160.052	+0.013	160.040	160.068	−0.003	160.040	160.125	−0.060	160.040	160.215	−0.150
	Min	160.000	160.003	−0.028	160.000	160.027	−0.052	160.000	160.043	−0.068	160.000	160.100	−0.125	160.000	160.190	−0.215
200	Max	200.046	2(°).033	+0.042	200.046	200.060	+0.015	200.046	200.079	−0.004	200.046	200.151	−0.076	200.046	200.0163	−0.190
	Min	200.000	200.004	−0.033	200.000	200.031	−0.060	200.000	200.050	−0.079	200.000	200.122	−0.151	200.000	200.236	−0.265
250	Max	250.046	250.033	+0.042	250.046	250.060	+0.015	250.046	250.079	−0.004	250.046	250.169	−0.094	250.046	250.313	−0.238
	Min	250.000	250.004	−0.033	250.000	250.031	−0.060	250.000	250.050	−0.079	250.000	250.140	−0.169	250.000	250.284	−0.313
300	Max	300.052	300.036	+0.048	300.052	300.066	+0.018	300.052	300.088	−0.004	300.052	300.202	−0.118	300.052	300.382	−0.298
	Min	300.000	300.004	−0.036	300.000	300.034	−0.066	300.000	300.056	−0.088	300.000	300.170	−0.102	300.000	300.350	−0.382
400	Max	400.057	400.040	0.053	400.057	400.073	+0.020	400.057	400.098	−0.005	400.057	400.244	−0.151	400.057	400.471	−0.378
	Min	400.000	400.004	−0.040	400.000	400.037	−0.073	400.000	400.062	−0.098	400.000	400.205	−0.244	400.000	400.435	−0.471
500	Max	500.063	500.045	+0.058	500.063	500.080	+0.023	500.063	500.108	−0.005	500.063	500.293	−0.189	500.063	500.580	−0.477
	Min	500.000	500.005	−0.045	500.000	500.040	−0.080	500.000	500.068	−0.108	500.000	500.252	−0.292	500.000	500.540	−0.580

All dimensions are in millimeters.

Preferred fits for other sizes can be calculated from data given in ANSI B4.2–1978 (R1984).

[†]A plus sign indicates clearance; a minus sign indicates interference.

Source: Reprinted courtesy of The American Society of Mechanical Engineers.

ANSI PREFERRED SHAFT BASIS METRIC CLEARANCE FITS–METRIC UNITS

American National Standard Preferred Shaft Basis Metric Clearance Fits (ANSI B4.2–1978, R1984)

Basic Size		Loose Running			Free Running			Close Running			Sliding			Locational Clearance		
		Hole C11	Shaft h11	Fit†	Hole D9	Shaft h9	Fit†	Hole F5	Shaft h7	Fit†	Hole G7	Shaft h6	Fit†	Hole H7	Shaft h6	Fit†
1	Max	1.120	1.000	0.180	1.045	1.000	0.070	1.020	1.000	0.030	1.012	1.000	0.018	1.010	1.000	0.016
	Min	1.060	0.940	0.060	1.020	0.975	0.020	1.006	0.990	0.006	1.002	0.994	0.002	1.000	0.994	0.000
1.2	Max	1.320	1.200	0.180	1.245	1.200	0.070	1.220	1.200	0.030	1.212	1.200	0.018	1.210	1.200	0.016
	Min	1.260	1.140	0.060	1.220	1.175	0.020	1.206	1.190	0.006	1.202	1.194	0.002	1.200	1.194	0.000
1.6	Max	1.720	1.600	0.180	1.645	1.600	0.070	1.620	1.600	0.030	1.612	1.600	0.018	1.610	1.600	0.016
	Min	1.660	1.540	0.060	1.620	1.575	0.020	1.606	1.590	0.006	1.602	1.594	0.002	1.600	1.594	0.000
2	Max	2.120	2.000	0.180	2.045	2.000	0.070	2.020	2.000	0.030	2.012	2.000	0.018	2.010	2.000	0.016
	Min	2.060	1.940	0.060	2.020	1.975	0.020	2.006	1.990	0.006	2.007	1.994	0.002	2.000	1.994	0.000
2.5	Max	2.620	2.500	0.180	2.545	2.500	0.070	2.520	2.500	0.030	2.512	2.500	0.018	2.510	2.500	0.016
	Min	2.560	2.440	0.060	2.520	2.475	0.020	2.506	2.490	0.006	2.502	2.494	0.002	2.500	2.494	0.000
3	Max	3.120	3.000	0.180	3.045	3.000	0.070	3.020	3.000	0.030	3.012	3.000	0.018	3.010	3.000	0.016
	Min	3.060	2.940	0.060	3.020	2.975	0.020	3.006	2.990	0.006	3.002	2.994	0.002	3.000	2.994	0.000
4	Max	4.145	4.000	0.220	4.060	4.000	0.090	4.028	4.000	0.040	4.016	4.000	0.024	4.012	4.000	0.020
	Min	4.070	3.925	0.070	4.030	3.970	0.030	4.010	3.988	0.010	4.004	3.992	0.004	4.000	3.992	0.000
5	Max	5.145	5.000	0.220	5.060	5.000	0.090	5.028	5.000	0.040	5.016	5.000	0.024	5.012	5.000	0.020
	Min	5.070	4.925	0.070	5.030	4.970	0.030	5.010	4.988	0.010	5.004	4.992	0.004	5.000	4.992	0.000
6	Max	6.145	6.000	0.220	6.060	6.000	0.090	6.028	6.000	0.040	6.016	6.000	0.024	6.012	6.000	0.020
	Min	6.070	5.925	0.070	6.030	5.970	0.030	6.010	5.988	0.010	6.004	5.992	0.004	6.000	5.992	0.000
8	Max	8.170	8.000	0.260	8.076	8.000	0.112	8.035	8.000	0.050	8.020	8.000	0.029	8.015	8.000	0.024
	Min	8.080	7.910	0.080	8.040	7.964	0.040	8.013	7.985	0.013	8.005	7.991	0.005	8.000	7.991	0.000
10	Max	10.170	10.000	0.260	10.076	10.000	0.112	10.035	10.000	0.050	10.020	10.000	0.029	10.015	10.000	0.024
	Min	10.080	9.910	0.080	10.040	9.964	0.040	10.013	9.985	0.013	10.005	9.991	0.005	10.000	9.991	0.000
12	Max	12.205	12.000	0.315	12.093	12.000	0.136	12.043	12.000	0.061	12.024	12.000	0.035	12.018	12.000	0.029
	Min	12.095	11.890	0.095	12.050	11.957	0.050	12.016	11.982	0.026	12.006	11.989	0.006	12.000	11.989	0.000
16	Max	16.205	16.000	0.315	16.093	16.000	0.136	16.043	16.000	0.061	16.024	16.000	0.035	16.018	16.000	0.029
	Min	16.095	15.890	0.095	16.050	15.957	0.050	16.016	15.982	0.016	16.006	15.989	0.006	16.000	15.989	0.000
20	Max	20.240	20.000	0.370	20.117	20.000	0.169	20.053	20.000	0.074	20.028	20.000	0.041	20.021	20.000	0.034
	Min	20.110	19.870	0.110	20.065	19.948	0.065	20.020	19.979	0.020	20.007	19.987	0.007	20.000	19.987	0.000

Basic Size		Loose Running			Free Running			Close Running			Sliding			Locational Clearance		
		Hole C11	Shaft h11	Fit[†]	Hole D9	Shaft h9	Fit[†]	Hole F8	Shaft h7	Fit[†]	Hole G7	Shaft h6	Fit[†]	Hole H7	Shaft h6	Fit[†]
25	Max	25.240	25.000	0.370	25.117	25.000	0.169	25.053	25.000	0.074	25.028	25.000	0.041	25.021	25.000	0.034
	Min	25.110	24.870	0.110	25.065	24.948	0.065	25.020	24.979	0.020	25.007	24.987	0.007	25.000	14.987	0.000
30	Max	30.240	30.000	0.370	30.117	30.000	0.169	30.053	30.000	0.074	30.028	30.000	0.041	30.021	30.000	0.034
	Min	30.110	29.870	0.110	30.065	29.948	0.065	30.020	29.979	0.020	30.007	29.987	0.007	30.000	29.987	0.000
40	Max	40.280	40.000	0.440	40.142	40.000	0.204	40.064	40.000	0.089	40.034	40.000	0.050	40.025	40.000	0.041
	Min	40.120	39.840	0.120	40.080	39.938	0.080	40.025	39.975	0.025	40.009	39.984	0.009	40.000	39.984	0.000
50	Max	50.290	50.000	0.450	50.142	50.000	0.204	50.064	50.000	0.089	50.034	50.000	0.050	50.025	50.000	0.041
	Min	50.130	49.840	0.130	50.080	49.938	0.080	50.025	49.975	0.025	50.009	49.984	0.009	50.000	49.984	0.000
60	Max	60.330	60.000	0.520	60.174	60.000	0.248	60.076	60.000	0.106	60.040	60.000	0.059	60.030	60.000	0.049
	Min	60.140	59.810	0.140	60.100	59.926	0.100	60.030	59.970	0.030	60.010	59.981	0.010	60.000	59.981	0.000
80	Max	80.340	80.000	0.530	80.174	80.000	0.248	80.076	80.000	0.106	80.040	80.000	0.059	80.030	80.000	0.049
	Min	80.150	79.810	0.150	80.100	79.926	0.100	80.030	79.970	0.030	80.010	79.981	0.010	80.000	79.981	0.000
100	Max	100.390	100.000	0.610	100.207	100.000	0.294	100.090	100.000	0.125	100.047	100.000	0.069	100.035	100.000	0.057
	Min	100.270	99.780	0.170	100.120	99.913	0.120	100.036	99.965	0.036	100.012	99.978	0.012	100.000	99.978	0.000
120	Max	120.400	120.000	0.620	120.207	120.000	0.294	120.090	120.000	0.125	120.047	120.000	0.069	120.035	120.000	0.057
	Min	120.180	119.780	0.180	120.120	119.923	0.120	120.036	119.965	0.036	120.012	219.978	0.012	120.000	119.978	0.000
160	Max	160.460	160.000	0.710	160.245	160.000	0.345	160.106	160.000	0.146	160.054	160.000	0.079	160.040	160.000	0.063
	Min	160.210	159.750	0.210	160.145	159.900	0.145	160.043	159.960	0.043	160.014	159.975	0.014	160.000	159.975	0.000
200	Max	200.530	200.000	0.820	200.285	200.000	0.400	200.122	200.000	0.168	200.061	200.000	0.090	200.046	200.000	0.075
	Min	200.240	199.710	0.240	200.170	199.885	0.170	200.050	199.954	0.050	200.015	199.971	0.015	200.000	199.911	0.000
250	Max	250.570	250.000	0.860	250.285	250.000	0.400	250.122	250.000	0.168	250.061	250.000	0.090	250.046	250.000	0.075
	Min	250.280	249.710	0.280	250.170	249.885	0.170	250.050	249.954	0.050	250.015	249.971	0.015	250.000	249.971	0.000
300	Max	300.650	300.000	0.970	300.320	300.000	0.450	300.137	300.000	0.189	300.069	300.000	0.101	300.052	300.000	0.084
	Min	300.330	299.680	0.330	300.190	299.870	0.190	300.056	299.948	0.056	300.017	299.968	0.017	300.000	299.968	0.000
400	Max	400.760	400.000	1.120	400.350	400.000	0.490	400.151	400.000	0.208	400.075	400.000	0.111	400.057	400.000	0.093
	Min	400.400	399.640	0.400	400.210	399.860	0.210	400.062	399.943	0.062	400.018	399.964	0.018	400.000	399.964	0.000
500	Max	500.880	500.000	1.280	500.385	500.000	0.540	500.165	500.000	0.228	500.083	500.000	0.123	500.063	500.000	0.103
	Min	500.480	499.600	0.480	500.230	499.845	0.230	500.068	499.937	0.068	500.020	499.960	0.020	500.000	499.960	0.000

All dimensions are in millimeters.

Preferred fits for other sizes can be calculated from data given in ANSI B4.2–1978 (R1984)

[†]All fits shown in this table have clearance.

Source: Reprinted courtesy of The American Society of Mechanical Engineers.

☐ ANSI PREFERRED SHAFT BASIS METRIC TRANSITION AND INTERFERENCE FITS–METRIC UNITS

American National Standard Preferred Shaft Basis Metric Transition and Interference Fits (ANSI B4.2–1978, R 1984)

Basic Size		Locational Transition			Locational Transition			Locational Interference			Medium Drive			Force		
		Hole K7	Shaft h6	Fit†	Hole N7	Shaft h6	Fit†	Hole P7	Shaft h6	Fit†	Hole S7	Shaft h6	Fit†	Hole U7	Shaft h6	Fit†
1	Max	1.000	1.000	+0.006	0.996	1.000	+0.002	0.994	1.000	0.000	0.986	1.000	-0.008	0.982	1.000	-0.012
	Min	0.990	0.994	-0.010	0.986	0.994	-0.014	0.984	0.994	-0.016	0.976	0.994	-0.024	0.972	0.994	-0.028
1.2	Max	1.200	1.200	+0.006	1.196	1.200	+0.002	1.194	1.200	0.000	1.186	1.200	-0.008	1.182	1.200	-0.012
	Min	1.190	1.194	-0.010	1.186	1.194	-0.014	1.184	1.194	-0.016	1.176	1.194	-0.024	1.172	1.194	-0.028
1.6	Max	1.600	1.600	+0.006	1.596	1.600	+0.002	1.594	1.600	0.000	1.586	1.600	-0.008	1.582	1.600	-0.012
	Min	1.590	1.594	-0.010	1.586	1.594	-0.014	1.584	1.594	-0.016	1.576	1.594	-0.024	1.572	1.594	-0.028
2	Max	2.000	2.000	+0.006	1.996	2.000	+0.002	1.994	2.000	0.000	1.986	2.000	-0.008	1.982	2.000	-0.012
	Min	1.990	1.994	-0.010	1.986	1.994	-0.014	1.984	1.994	-0.016	1.976	1.994	-0.024	1.972	1.994	-0.028
2.5	Max	2.500	2.500	+0.006	2.496	2.500	+0.002	2.494	2.500	0.000	2.486	2.500	-0.008	2.482	2.500	-0.012
	Min	2.490	2.494	-0.010	2.485	2.494	-0.024	2.484	2.494	-0.016	2.476	2.494	-0.024	2.472	2.494	-0.028
3	Max	3.000	3.000	+0.006	2.996	3.000	+0.002	2.994	3.000	0.000	2.986	3.000	-0.008	2.982	3.000	-0.012
	Min	2.990	2.994	-0.010	2.986	2.994	-0.014	2.984	2.994	-0.016	2.976	2.994	-0.024	2.972	2.994	-0.028
4	Max	4.003	4.000	+0.011	3.996	4.000	+0.004	3.992	4.000	0.000	3.985	4.000	-0.007	3.981	4.000	-0.011
	Min	3.991	3.992	-0.009	3.984	3.992	-0.016	3.980	3.992	-0.020	3.973	3.992	-0.027	3.969	3.992	-0.031
5	Max	5.003	5.000	+0.011	4.996	5.000	+0.004	4.992	5.000	0.000	4.985	5.000	-0.007	4.981	5.000	-0.011
	Min	4.991	4.992	-0.009	4.984	4.992	-0.016	4.980	4.992	-0.020	4.973	4.992	-0.027	4.969	4.992	-0.031
6	Max	6.003	6.000	+0.011	5.996	6.000	+0.004	5.992	6.000	0.000	5.985	6.000	-0.007	5.981	6.000	-0.011
	Min	5.991	5.992	-0.009	5.984	5.992	-0.016	5.980	5.992	-0.020	5.973	5.992	-0.027	5.969	5.992	-0.031
8	Max	8.005	8.000	+0.014	7.996	8.000	+0.005	7.992	8.000	0.000	7.983	8.000	-0.008	7.978	8.000	-0.013
	Min	7.990	7.991	-0.010	7.981	7.991	-0.019	7.976	7.991	-0.024	7.968	7.991	-0.032	7.963	7.991	-0.037
10	Max	10.005	10.000	+0.014	9.996	10.000	+0.005	9.991	10.000	0.000	9.983	10.000	-0.008	9.978	10.000	-0.013
	Min	9.990	9.991	-0.010	9.981	9.991	-0.019	9.976	9.991	-0.024	9.968	9.991	-0.032	9.963	9.991	-0.037
12	Max	12.006	12.000	+0.017	11.995	12.000	+0.006	11.989	12.000	0.000	11.979	12.000	-0.010	11.974	12.000	-0.015
	Min	11.988	11.989	-0.012	11.977	11.989	-0.023	11.971	11.989	-0.029	11.961	11.989	-0.039	11.956	11.989	-0.044
16	Max	16.006	16.000	+0.017	15.995	16.000	+0.006	15.989	16.000	0.000	15.979	16.000	-0.010	15.974	16.000	-0.015
	Min	15.988	15.989	-0.012	15.977	15.989	-0.023	15.971	15.989	-0.019	15.961	15.989	-0.039	15.956	15.989	-0.044
20	Max	20.006	20.000	+0.019	19.993	20.000	+0.006	19.986	20.000	-0.001	19.973	20.000	-0.014	19.967	20.000	-0.020
	Min	19.985	19.987	-0.015	19.972	19.987	-0.028	19.965	19.987	-0.035	19.952	19.987	-0.045	19.946	19.987	-0.054

Basic Size		Locational Transition			Locational Transition			Locational Interference			Medium Drive			Force		
		Hole K7	Shaft h6	Fit†	Hole N7	Shaft h6	Fit†	Hole P7	Shaft h6	Fit†	Hole S7	Shaft h6	Fit†	Hole U7	Shaft h6	Fit†
25	Max	25.006	25.000	+0.019	24.993	25.000	+0.006	24.986	25.000	-0.001	24.973	25.000	-0.014	24.960	25.000	-0.027
	Min	24.985	24.987	-0.015	24.972	24.987	-0.028	24.963	24.987	-0.035	24.952	24.987	-0.048	24.939	24.987	-0.061
30	Max	30.006	30.000	+0.019	29.993	30.000	+0.006	29.986	30.000	-0.001	29.973	30.000	-0.014	29.960	30.000	-0.027
	Min	29.985	29.987	-0.015	29.972	29.987	-0.028	29.965	29.987	-0.035	29.952	29.987	-0.048	29.939	29.987	-0.061
40	Max	40.007	40.000	+0.023	39.992	40.000	+0.008	39.983	40.000	-0.001	39.966	40.000	-0.018	39.949	40.000	-0.035
	Min	39.982	39.984	-0.018	39.967	39.984	-0.033	39.958	39.984	-0.042	39.941	39.984	-0.059	39.924	39.984	-0.076
50	Max	50.007	50.000	+0.023	49.992	50.000	+0.008	49.983	50.000	-0.001	49.966	50.000	-0.018	49.939	50.000	-0.055
	Min	49.982	49.984	-0.018	49.967	49.984	-0.033	49.958	49.984	-0.042	49.941	49.984	-0.059	49.914	49.984	-0.086
60	Max	60.009	60.000	+0.028	59.991	60.000	+0.010	59.979	60.000	-0.002	59.958	60.000	-0.023	59.924	60.000	-0.087
	Min	59.979	59.981	-0.021	59.961	59.981	-0.039	59.949	59.981	-0.051	59.928	59.981	-0.072	59.894	59.981	-0.106
80	Max	80.009	80.000	+0.028	79.991	80.000	+0.010	79.979	80.000	-0.002	79.952	80.000	-0.029	79.909	80.000	-0.072
	Min	79.979	79.981	-0.021	79.961	79.981	-0.039	79.949	79.981	-0.051	79.922	79.981	-0.078	79.879	79.981	-0.121
100	Max	100.010	100.000	+0.032	99.990	100.000	+0.012	99.976	100.000	-0.002	99.942	100.000	-0.036	99.889	100.000	-0.089
	Min	99.975	99.978	-0.025	99.955	99.978	-0.045	99.941	99.978	-0.059	99.907	99.978	-0.093	99.854	99.978	-0.146
120	Max	120.010	120.000	+0.032	119.990	120.000	+0.012	119.976	120.000	-0.002	119.934	120.000	-0.044	119.869	120.000	-0.109
	Min	119.975	119.978	-0.025	119.955	119.978	-0.045	119.941	119.978	-0.059	119.899	119.978	-0.101	119.834	119.978	-0.166
160	Max	160.012	160.000	+0.037	159.988	160.000	+0.013	159.972	160.000	-0.003	159.915	160.000	-0.060	159.825	160.000	-0.150
	Min	159.972	159.975	-0.028	159.948	159.975	-0.053	159.932	159.975	-0.068	159.875	159.975	-0.125	159.785	159.975	-0.213
200	Max	200.013	200.000	+0.042	199.986	200.000	+0.015	199.967	200.000	-0.004	199.895	200.000	-0.076	199.781	200.000	-0.190
	Min	199.967	199.971	-0.033	199.940	199.971	-0.060	199.921	199.971	-0.079	199.849	199.971	-0.151	199.735	199.971	-0.265
250	Max	250.013	250.000	+0.042	249.986	250.000	+0.015	249.967	250.000	-0.004	249.877	250.000	-0.094	249.733	250.000	-0.238
	Min	249.967	249.971	-0.033	249.940	249.971	-0.060	249.921	249.971	-0.079	249.831	249.971	-0.169	249.687	249.971	-0.313
300	Max	300.016	300.000	+0.048	299.986	300.000	+0.018	299.964	300.000	-0.004	299.850	300.000	-0.118	299.670	300.000	-0.298
	Min	299.964	299.968	-0.036	299.934	299.968	-0.066	299.912	299.968	-0.088	299.798	299.968	-0.202	299.618	299.968	-0.382
400	Max	400.017	400.000	+0.053	399.984	400.000	+0.020	399.959	400.000	-0.005	399.813	400.000	-0.151	399.586	400.000	-0.378
	Min	399.960	399.964	-0.040	399.927	399.964	-0.073	399.902	399.964	-0.098	399.756	399.964	-0.244	399.529	399.964	-0.471
500	Max	500.018	500.000	+0.058	499.983	500.000	+0.023	499.955	500.000	-0.005	499.771	500.000	-0.189	499.483	500.000	-0.477
	Min	499.955	499.960	-0.045	499.920	499.960	-0.080	499.892	499.960	-0.108	499.708	499.960	-0.292	499.420	499.960	-0.580

All dimensions are in millimeters.

Preferred fits for other sizes can be calculated from data given in ANSI B4.2–1978 (R1984).

†A plus sign indicates clearance; a minus sign indicates interference.

Source: Reprinted courtesy of The American Society of Mechanical Engineers.

INDEX

Dwg 3-1

Dwg 3-2

Dwg 3-3

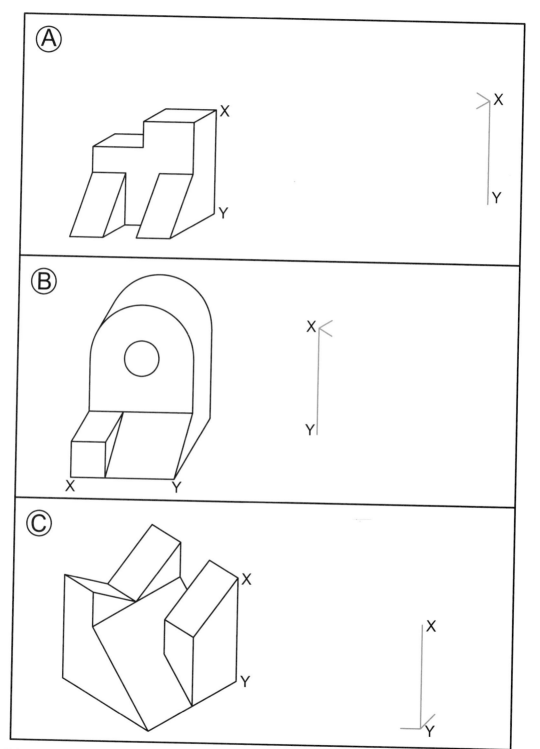

Ⓐ

X

X

Y

Y

Ⓑ

X

X

Y

Y

X

Y

Ⓒ

X

X

Y

Y

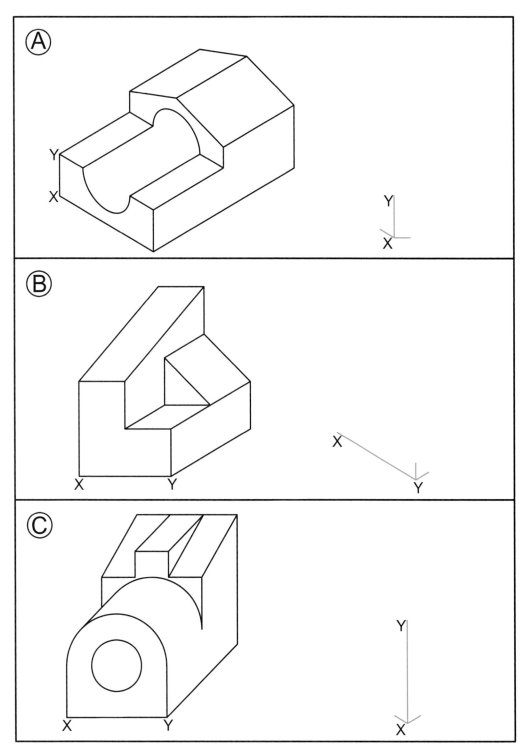

Dwg 4-2

Dwg 5-1

Dwg 5-2

VP ○

C ————————————— D

A ————————————— B

Dwg A-1

VPR

VPL

C _____ D

A _____ B

Dwg A-2